Bioethics in Our World

A Reader

FIRST EDITION

Bioethics in Our World

A Reader

Edited by Michael French

cognella®
SAN DIEGO

Bassim Hamadeh, CEO and Publisher

John Remington, Executive Editor

Anne Jones, Project Editor

Casey Hands, Production Editor

Emely Villavicencio, Senior Graphic Designer

Trey Soto, Licensing Specialist

Natalie Piccotti, Director of Marketing

Kassie Graves, Vice President of Editorial

Jamie Giganti, Director of Academic Publishing

3970 Sorrento Valley Blvd., Ste. 500, San Diego, CA 92121

CONTENTS

INTRODUCTION TO THE BOOK

I write this introduction from my apartment. I look out the window, and the once busy streets are nearly empty. The few people who are walking down the sidewalks are wearing cloth masks that cover their mouths and noses. The outbreak of SARS-COVID-19 discomforted us, changed us, and affected our relationships with medicine, science, and each other. As SARS-COVID-19 grew into a pandemic, we remained home, wore masks when leaving, limited contact with others, and kept on our minds the responsibility we had toward the health of others. Why do we do this? How are the regulations and choices that keep us indoors or masked justifiable? How can we balance the autonomy we hold as a defining virtue with the need for us all to lower the risk of harms?

Bioethics encapsulates ethical issues within the fields of medicine, health care, and research. We will explore issues from public health, psychiatry, genetics, and more as we examine the moral worth of actions within these fields. While we will spend some time examining theories, the purpose of this book is to raise and attempt to answer ethical questions—specifically, questions we may face as clinicians, researchers, and citizens. As "social distancing" was added to our lexicon and birthdays were spent on video-chats, we each have experienced our role in the SARS-COVID-19 pandemic. While we will not be as personally involved in all the cases in the book, the cases we will examine are about research conducted and medical experiences in hopes of finding healthier lives for all.

To tell a quick history of bioethics, I would like to use three examples of bioethics in practice. The Hippocratic Oath was a requirement for doctors in the Hippocratic tradition. It is often attributed to the philosopher and physician Hippocrates (or his followers), and the oath is estimated to be 2,400 years old. The oath, taken in the name of the gods Apollo, Asclepius, Hygeia, and Panacea, required doctors to adhere to a strict code of ethics and medicine. Doctors agreed to teach other brothers the art of medicine; not to euthanize or give a patient a deadly drug; not to give pessary to cause abortion; not to have sex with your patients, their families, or their slaves; and not to use the knife on your patients, among other requirements. This code of medical ethics came with devastating consequences for unethical actions. If the doctors executed this oath to the fullest of their ability, they would gain in reputation with men for all time to come, and if they failed, the opposite would be their lot. A modernized version is recited by graduating medical students.

In 1961, a committee was formed to decide which patients would receive the Scribner shunt, which allowed for outpatient hemodialysis for the first time. As a new technology, it was expensive for patients. There were far more individuals whose lives could be saved by the Scribner shunt than the number of shunts available. There had to be a way of deciding who would receive them and who would not. The executive committee decided that seven unpaid and anonymous community members would make the decision. Though the committee's proper title was the Admissions and Policies Committee of the Seattle Artificial Kidney Center at Swedish Hospital, it was labeled "The God Committee" in an exposé by journalist Shana Alexander in *LIFE* magazine. The committee was made up of a lawyer, a minister, a banker, a housewife, an official of state government, a labor leader, and a surgeon. Though they were given no direct instructions on how to choose recipients, they weighed a number of factors, including age and sex of the patient, marital status, number of dependents, net worth, emotional stability, educational background, nature of occupation, past performance and future potential, and references. After their decision-making was revealed by Alexander's work, the committee was disbanded.

Henry K. Beecher was an anesthesiologist and professor of medicine at Harvard Medical School. He was the author of "Ethics and Clinical Research" in the *New England Journal of Medicine* in 1966. Beecher wrote about experiments on humans of which he was aware. Among his 22 examples of ethically objectionable studies was one describing transplanting a melanoma from a woman to her mother, both of whom died of metastatic melanoma, and lowering patients' blood pressure to find out if the central nervous system or heart would collapse first. Among his ethical stances, Beecher decreed that informed consent could not be met if all expected and possible outcomes were not known by those giving consent and safety measures ought to be used if it is possible.

These examples demonstrate the necessity and ubiquity of bioethics. The influence of ethics aids and alters the decisions made in medicine, public health, and research. Throughout this book, we will examine cases and ethical theories as applicable. There have been a variety of ethical foundations utilized in the decision-making process, and it will be our responsibility to investigate and apply those theories to understand how best to use them. One common ethical theory utilized in bioethics is principlism. By following principles, one can act more ethically. By settling on universal principles, unforeseen choices can still be accounted for so long as the actions reflect the principles. In *Principles of Biomedical Ethics* (1979), Tom L. Beauchamp and James F. Childress developed four principles on which bioethics has focused. The four principles are autonomy (self-determination), beneficence (to do good), nonmaleficence (to do no harm), and justice. Beauchamp and Childress propose that by making choices that align with these principles, actions will be ethical. These principles are commonly used in bioethics but raise questions as well. For example, what is harm or what should be done when two principles conflict? We will return to these principles throughout the book, utilizing and questioning them. While theories offer foundation, critical thinking will be required for decision-making for real world problems.

Lives are lost and saved by decisions. Groups are persecuted, biases propagated, and unity built because of decisions. Harms are caused or prevented, voices are heard or silenced, and practitioners are supported or ignored based on the decisions we make. Ethics helps us make better decisions. Our purpose is to understand what we do, should do, and should strive to be.

How the Book Is Organized

This book is organized in eight units. The first unit examines four theories we shall utilize throughout the book. The next seven units are built around topics that are salient to understanding the nature of bioethics and the world in which bioethics exists. After each unit are discussion questions and recommended cases.

Unit I, "Theory," provides a foundation in ethical theories. This will allow us to apply them to the topics in this book and beyond. The unit presents the theories of utilitarianism, deontology, virtue ethics, and care ethics.

Unit II, "Research," focuses on the ethical issues in medical and scientific research. Individuals upon whom research is being done are in a vulnerable position. Their position requires trust in the researcher, and they are susceptible to various effects from the research done on them. We shall examine historical failures of research as well as ways we can research more ethically.

Unit III, "Eugenics," examines the history of eugenics and its relationship to some potentially eugenic practices today. Eugenics programs have taken place globally, with the United States having prominent eugenic programs. One contemporary program with a connection to eugenics we will examine is selective termination.

Unit IV, "Reproduction," investigates reproductive rights as well as the uses of various reproductive technologies. We will examine historical, contemporary, and potential reproduction practices to understand the relationship between ourselves and potential generations to come.

Unit V, "Experiences," demonstrates the variety of experiences that are faced by patients, researchers, and health care professionals. We will see the ways that experiences have varied based on race, gender, age, and ability and how those experiences result from a history of bias and stereotyping.

Unit VI, "Euthanasia and Physician-Assisted Suicide," examines the two titular practices. The role of the practitioner, the availability of the practices, and the safety of all persons involved are central issues in this unit. Bringing on our death has been a topic for millennia, and we shall see the contemporary arguments regarding its ethical status.

Unit VII, "Genes and Data," introduces us to stem cell research and gene-editing technology. Attempts to use these technologies appropriately become difficult, as there is not a global governing body for research. How we develop and utilize these technologies may change the history to come.

Unit VIII, "Medicalization," raises the question "What is an illness?" By medicalizing certain behaviors, society can increase or decrease support. Additionally, by medicalizing things that are not necessary to make medical conditions, such as an individual's sexuality, it causes a stigma that harms individuals who would be deemed abnormal.

UNIT I

Theory

....

Unit I Introduction

Outrage is easy. It does not take much work for us to think of something that causes outrage to rise within us. In a world rife with unethical action, it is easy to feel outrage but more difficult to understand what it is about such actions that cause us to feel the anger and disgust rising. Justifying and understanding outrage is more difficult. This chapter will relate to outrage in two ways: it will allow you to better understand the cause and allow you to question the justification for such outrage.

In this unit, we will study theories that underlie evaluations. The four theories examined in this unit are utilitarianism, deontology, virtue ethics, and care ethics. These are not the only ethical theories utilized in Bioethics (for instance, principlism which was referenced in the introduction to this book), but these offer a range of methods by which we can examine actions, choices, and cases.

"A Note on Utilitarianism" will introduce the titular theory, which was developed by Jeremy Bentham and John Stuart Mill with the purpose of maximizing utility. In this theory, ethical actions are understood by their consequences, the outcomes of the action in question. The goodness of the action is understood by how it increases utility, an anachronistic word commonly updated to "well-being" or "happiness."

"A Note on Deontology" focuses on deontology, an ethical theory popularized by Immanuel Kant. In deontology, one's duty and moral motivation for an action is the determinant of moral value. These moral rules must be universalizable and treat individuals as ends in themselves.

Virtue ethics focuses on the best qualities of an individual, a role, or a society to understand how to best improve upon those traits. Ethical actions are those that develop virtues. For instance, a virtue ethicist may examine the role of the doctor and ask "What are the most important virtues for a doctor?" to clarify how a doctor should act. This theory is presented in "Selections from Aristotle."

The final reading for this unit is the introduction from the book *Care, Autonomy, and Justice: Feminism and the Ethic of Care.* In this selection, we will examine the ethics of care, a theory focusing on promoting relationships, nurturing trust, and cooperation. Ethics of care focuses on the bonds between us and our relationships as a foundation for moral value.

These theories may not guarantee ethical decisions will be made by following them. Theories alone are often insufficient to understanding the ethical action to take. These are tools to reframe and enhance our ethical thinking. Each of these may have drawbacks, but only by examining useful instances and flawed uses of each can we hone our abilities to better understand how to act.

A Note on Deontology

R. Edward Freeman, Patricia H. Werhane, and Scott B. Sonenshein

O ne branch of thinking about ethics focuses on the rightness of an action or the motives of the decision maker. Deontologists argue that an act is morally right if it is done out of a sense of duty. The "father" of this point of view is the eighteenth-century German philosopher Immanuel Kant. Kant argued that ethics and ethical reasoning involves the uniquely human ability of making rational, free choices from a number of possible alternatives. Moreover, Kant claimed that moral decisions have to do with what we can bring under our control—our choices and our intentional actions rather than the consequences of our actions, which we cannot always regulate. According to Kant, moral choices and moral actions also have to do with actions that apply to all human beings. The best moral choices are those that you would want others to do, even when you cannot make yourself an exception. Moral choices also respect others as persons. A moral act, then, can be defined as an act done from a sense of principle, a principle that you would want others to adopt in their actions, and an act that respects others as moral agents. In contemporary terms, a moral act would be an act that reasonable persons would agree was right, that you would expect others to do, that could be embodied into a universally applicable law, and that respects people equally.

From a Kantian perspective, we might ask the following kinds of questions when deciding what action to take:

- Does the action set positive or negative precedents?
- Is it an action that is acceptable to other reasonable persons?
- Is it "legislatible": that is, is it applicable to other similar situations?
- Does it respect or at least not denigrate human dignity?

Kant's analysis introduces us to a criterion for moral assessment that is not simply an evaluation of utility or foreseeable outcomes nor does it merely reflect law, codes, or societal mores. Some actions are judged to be wrong not because of their positive or negative outcomes. Rather, actions are judged to be wrong because they violate standards for acceptable behavior or moral rules. For example, most of us would agree that we should avoid deception, keep promises, respect basic rights of everyone equally (e.g., the rights to life, freedom, and privacy), treat people fairly, and avoid causing harm. These rules—the standards of truth telling, promise keeping, equal rights, fairness, and not causing suffering—are part of the ordinary morality that reasonable people would agree should hold for everyone. Reasonable people advocate and defend these standards (and others) because it is in the interests of all of us if these rules are followed, and it is usually not in our long-term interests if these rules are habitually violated. Moral rules set the criteria for acceptable behavior by specifying how all are expected to act, without making an exception for ourselves.

Kant argued that moral rules do not have exceptions. For example, consider his view on lying. Even if one had good reason to lie—say, protecting a friend from physical torture—Kant would still argue that it is wrong to lie. Unlike the utilitarian who would argue that the lie is justified because of the consequences, Kant would argue that no outcome could excuse breaking a moral law. However, most of us think that moral standards have exceptions; there are times when one cannot respect all moral rules or respect them equally. For instance, when one's life is threatened, one often must kill in self-defense. However, one can override a rule or standard (e.g., that everyone has a right to life) only when one has good reasons, reasons that other reasonable people would accept as being sound (e.g., self-defense) because they appeal to another standard (e.g., equal rights include *my* right to life and freedom).[1] Moreover, although moral rules are general rules, how they are interpreted depends, in part, on the context of a particular situation. For example, one would not be required *never* to lie, even in extreme circumstances such as national emergencies or war.

Because they are general standards, moral rules help us to evaluate laws, codes, and social practices. But where do moral rules come from? There are a number of tempting explanations. Perhaps moral rules are God-given or part of human nature. Proving this involves us in religious debates that would take us afield. Another way to derive moral rules is through a thought experiment. Imagine that you and your family and colleagues are on a spaceship heading for a distant planet you are going to populate. You are not sure how long the journey will take or what the conditions will be when you arrive. During the journey you and the other passengers must set up the guidelines for behavior and the standards for law and justice that will regulate the new colony. Since you do not know in advance what your status will be on the new planet, you are likely to choose guidelines that will give fair opportunities for all settlers, including yourself and future generations. Under those conditions, you are likely to set out guidelines that specify equal rights to life and basic liberties, protect contracts and promises, punish lying, stealing, cheating, and murder, and provide everyone with an equal opportunity to achieve well-being. Assuming that the spaceship travelers are reasonable people, the guidelines or rules you develop will be those agreed upon by all members of the party. One can think of moral rules, then,

---

1 Adam Smith, *Lectures on Jurisprudence* (Oxford: Oxford University Press, 1978), especially A (i.ii.12–14); and Bernard Gert, *Morality* (New York: Oxford University Press, 1988), especially chapters 1–7.

as those rules reasonable people, in ignorance of their own circumstances or future, would agree are the best standards for their own behavior and the behavior of others.

Some commonly held moral rules include:

- non-malevolence (avoid harming others),
- keep promises, honor contracts,
- respect persons and their rights, and respect property (however defined),
- mutual aid (for one's family, and/or community, and/or nation, and/or world).

A Note on Utilitarianism

R. Edward Freeman, Patricia H. Werhane, and Scott B. Sonenshein

O ne main branch of thinking about ethics focuses on what is good. Such teleological the-ories claim that morality involves maximizing "the good," though there is a great deal of disagreement over precisely what constitutes the good. For some, the good is power, knowledge, or perfection. Others identify the good with pleasure or happiness and are known as "utilitarians." Utilitarians believe that actions which are ethical must maximize the amount of happiness or utility in a society.[1]

Whenever one analyzes an ethical issue in business, a chief concern is with the costs and benefits of the alternatives. Although such an approach appears to be primarily economic, it is buttressed by a well-known ethical theory, utilitarianism. Utilitarianism links what is universally desired with what is desirable and argues that what is most important and universally valued is the satisfaction of desires or interests, human pleasure or happiness, or the reduction of human suffering. A utilitarian judges human action in terms of its outcomes; that is, she measures the positive or negative utility of an action itself, not merely what it was meant to achieve. The best sort of decision, then, is one that maximizes human interests, best satisfies desires or pleasures, or minimizes harm. (Human interests include life, health, wealth, human dignity, autonomy, or pleasure.)

A utilitarian measures harms and benefits in terms of their qualitative and quantitative merit, long-term and short-term results, or immediate or latent satisfaction. In making a moral decision, one takes into account how a particular action or set of decisions would affect the greatest number of people, impartially evaluating each person's interests.[2] The best kind of moral decision making, then, impartially applies the principle of utility over the range of persons affected by the decision or its outcomes. It weighs each person and his or her interests equally and seeks an equitable, if not an equal, distribution of benefits or harms. The best outcome maximizes the interests (or contributes

---

1 For further discussion, see William Frankena, *Ethics* (Englewood Cliffs, NJ: Prentice Hall, 1973), especially chapter 2.

2 As John Stuart Mill says, "As between his [the agent's] own happiness and that of others, utilitarianism requires him to be as strictly impartial as a disinterested and benevolent spectator," in *Utilitarianism* (Cleveland: Meridian Books, 1968), 268. Mill also quotes Bentham as saying, "Everybody to count for one, nobody for more than one" (page 319).

to the happiness) of the greatest number, leading to more benefits than harms for most people. At a minimum, utilitarian morality demands reducing harm, all things considered.

Some utilitarians argue that one ought to judge each act separately according to its utility. So-called *act utilitarians* seek to maximize the good for each particular action. For example, if a doctor with limited resources could only perform one surgery for two dying patients, she would ask herself which treatment would bring the most happiness. However, in some cases, act utilitarianism requires apparently immoral acts. Consider Ursula Le Guin's story, "The Ones Who Walk Away from Omelas." Omelas is a utopia—full of happiness and filled with culture, prosperity, entertainment, and beauty. However, Omelas's good fortune is conditional on the pain and suffering of one child who is enslaved in a small, windowless room.

> They all know it is there, all the people of Omelas. Some of them have come to see it, others are content merely to know it is there. They all know that it has to be there. Some of them understand why, and some do not, but they all understand that their happiness, the beauty of their city, the tenderness of their friendships, the health of their children, the wisdom of their scholars, the skill of their makers, even the abundance of their harvest and the kindly weather of their skies, depend wholly on this child's abominable misery.[3]

The act utilitarian has to consider the possibility that Omelas is a just society. One could hardly think of a community that produces more happiness than Omelas—even after factoring in the unhappiness of the enslaved child.

To account for such morally bizarre outcomes, *rule utilitarianism* emerged as a new approach to ethics. Instead of judging each particular act, the rule utilitarian searches for general "rules" that she thinks will bring the most happiness to society. For example, the rule utilitarian might claim that "always telling the truth" will bring the greatest good to society. This rule then governs the utilitarian's ethical choices, even if breaking the rule in a particular case (e.g., telling a white lie) will bring more happiness than following the rule in a particular case.

Making choices based on utility is a very basic, normative decision-making practice, but it can be tricky. It could be the case that some sorts of practices are wrong for other reasons, despite their widespread practice or utility. The famous nineteenth-century utilitarian John Stuart Mill argued that freedom is a unique and primary interest of human beings and that other interests and benefits are subsidiary to freedom, including economic welfare. In other words, actions that appear to maximize utility really don't if basic freedoms are violated or not realized.[4]

---

3 Ursula Le Guin, "The Ones Who Walk Away from Omelas," in *The Wind's Twelve Quarters* (New York: Harper, 1975), 345–57.

4 John Stuart Mill, *Utilitarianism,* especially chapter 5.

Utilitarianism raises some important considerations that must be accounted for in any method of ethical decision making:

- any decision must be impartial, treating each person equally but no more than equally;
- ideally, no decision is acceptable that increases harms of any sort, even to a small number of people; more practically, large or important harms should be avoided;
- an ideal decision is one that maximizes the pleasures, preferences, desires, interests, or well-being of the greatest number; and,
- all consequences of a decision, economic and noneconomic, must be taken into account.

Aristotle

Charles Cardwell

....

The Good

In the last chapter we discussed the predicate "is good." The word "good" can also be used as a noun or as an adjective. The meaning of the noun flows from that of the predicate. An object (or state of affairs) that satisfies (someone's) desire(s) is called "a good." We use this term commonly when we speak of merchants selling their goods.

Aristotle opens his *Nicomachean Ethics* by observing that goods may be thought of as the ends (outcome, conclusion, product) of art, inquiry, action, or pursuit. Thus, for example, a shoe would be the end of the cobbler's art. As there are many arts, inquiries, actions, and pursuits, there will be many ends. Even within a particular endeavor, there may be multiple ends, some of which may be subsidiary to others. Carpenters may, for example, build scaffolding so that masons can lay bricks for a weatherproof exterior to a dwelling being built as a home for a pair of newlyweds. The "chief good" here, the one for the sake of which the other goods are pursued, is the newlyweds' home; all the other goods are subsidiary to it.

What does this have to do with ETHICS?

If there is some end at which all persons aim, an end achievable by human action, valued for itself, and to which all other goods are subsidiary, Aristotle observes, this must be the chief good, *the highest purpose or function for man.* Let's call this "THE GOOD." To attain THE GOOD would be to realize a good life.

What is THE GOOD (for man) … or is there no such good? Aristotle finds universal agreement that **eudaimonia**—a state of well-being most commonly rendered in English as "happiness"—is such a good. However, he finds no universal agreement as to exactly what happiness is. Some identify happiness with pleasure, but beasts seem capable of experiencing pleasure and so this cannot be THE GOOD for *man.* Some identify happiness with power or honor, but this won't do either because power and honor are bestowed by others and hence ultimately depend upon the actions of others, not of oneself. Some identify happiness with virtue, but virtue is compatible with total inactivity, with being asleep, even with the greatest suffering and misfortune; virtue cannot be THE GOOD. Some identify happiness with wealth; however, wealth is not desired for its own sake but rather because it is useful for the sake of something else. Other proposals seem even weaker candidates, because they change with changing conditions. A person who is ill, for example, may identify health with happiness, but when well, that same person would offer a different candidate.

Finding only negative results with this approach, Aristotle tries a different tack. The word "good" can appear not only as a predicate or a noun but also as an adjective. Aristotle points out that this usage is tied to **function:** "good" applies when something excels at carrying out its function. Thus, for example, the function of a knife is to cut. A *good* knife cuts cleanly and easily, is nicely balanced, comfortable to hold, takes a sharp edge, and holds its edge well. That is, a good knife is a knife that performs its function well.

Importantly, and in contrast to the subjective element in the predicate and noun, as an adjective, "good" carries a measure of objectivity. If one knows what the function of a thing is, then one can judge with considerable objectivity whether that thing excels in the performance of its function or not. Now, a thing's function, if we think about it, really amounts to the good—here "good" is a noun—for which it exists. Hence, if we knew the essential distinguishing function of man, we would know objectively what it would be to fulfill that function well, and consequently we would know objectively what a good life would be.

What then is the function of man ... the *essence* that is unique to the nature of man and which distinguishes man from all other species, indeed from all else? Other creatures share nutrition, growth, and perception with man. Only man, Aristotle suggests, is capable of what we might call **practical reason.**[1]

Practical (as opposed to theoretical or "pure") reason comprehends reflective thought as it pertains to voluntary decision and action. A familiar illustration comes from the story of Solomon and the two women arguing over a baby. You will remember that Solomon proposed splitting the baby in two and giving one half to each woman. By this ruse, Solomon identified the true mother who was aghast at this proposal and begged him to give the baby to the other woman instead. This is practical reason at work.

More needs to be said about exactly what practical reason is. But even without details, it seems apparent that practical reason is an essential characteristic that distinguishes man from all other beings. If so, Aristotle contends, THE GOOD (for man) must be **action in accord with practical reason,** and a good life would be one that accords with using practical reason well. [...].

---

1 The Greek *phronesis* is usually translated as "wisdom," "practical wisdom," or "prudence," but as Aristotle's notion is of a faculty, "reason" seems more appropriate than "wisdom." "Practical reason" is often used in translations of Kant, so it is important to be aware that the two authors are talking about two distinct notions even though the same language appears in translation.

Golden Mean

How does one get to morality from Aristotle's notion that THE GOOD is action in accord with practical reason?

Aristotle elaborates his notion of practical reason by use of what he calls the "Golden Mean."

What is this Golden Mean? It is not a mathematical notion, like median, mode, arithmetic, or geometric mean. It is rather the *right* amount—that is, the amount that is *neither too much nor too little*—and this amount *may* differ among times and individuals. The Golden Mean is perhaps most easily illustrated by the story of Goldilocks and the Three Bears. You will recall that for Goldilocks, one bed was too soft, another too hard, but one was just right; one chair was too big, another too small, but one was just right; and one bowl of porridge was too hot, another too cold, but one was just right. This notion of being "just right" is the notion of the Golden Mean. Presumably the hard bed would be at the Golden Mean for Papa Bear, the tiny chair at the Golden Mean for Baby Bear, and so on. The Golden Mean is the respective amount that is right for each individual. For Aristotle, the task of practical reason is to recognize the Golden Mean in all action. **Practical reason**, then, is just the ability to recognize whether something hits the Golden Mean, and if not, to recognize whether it falls on the side of excess or of deficiency.[2]

Aristotle sees in the Golden Mean a key to understanding **moral virtue.** He notices that both emotions and actions can be either deficient or excessive. Excessive fear combined with a deficit in confidence yields cowardice. A deficit in fear combined with excessive confidence yields foolhardiness. But when the right amount of fear combines with the right amount of confidence, we find courage. Cowardice and foolhardiness are vices; courage is a virtue. As a result of considering a number of vices and virtues, Aristotle contends that the case of courage is typical: each moral virtue represents a Golden Mean between vices of excess and vices of deficiency.

Moral virtue, then, is a state of character concerned with choice, lying at a mean, in this case, the mean relative to us, this being determined by a rational principle, that by which the man of practical wisdom would determine it.

It should be added that not all virtues are moral virtues. Some virtues are intellectual. How do moral and **intellectual virtues** differ?

Suppose students who know nothing about music sign up to play in the school band. Now imagine that the instructor teaches the students all about the staff, signature, note names, measures, the various kinds of notes, the "count" each indicates, and so on. After the students pass the test on this, the band instructor hands out flutes and trumpets and clarinets and trombones and so forth, and tells the students to play a song. Of course they cannot do so. To play an instrument, one must practice. Such practice involves trial and error *and* recognition whether the trial has hit its target or not. If on target, one tries to reproduce whatever produced that result; if off target, one tries again. In this

---

2 The Golden Mean need not vary among individuals. To sing middle C, a Golden Mean, one must be able to recognize both excess and deficiency in pitch and adjust one's voice accordingly.

sense, one cannot be taught to play an instrument. One can be guided, but only experimentation and practice will do the job.

Intellectual virtues, Aristotle tells us, are like the note names; they can be taught. Moral virtues, however, are like playing an instrument: that cannot be taught; it must be discovered and practiced. One becomes courageous by practicing courage—that is, by trying to find the proper balance of fear and confidence. Because the Golden Mean is only one point on an infinite continuum, hitting the Golden Mean will always be difficult.

Aristotle also advises us that, knowing we will likely miss the Golden Mean, we should always try to make our error on the side of the lesser vice. [...].

"Do Ethics" and "Be Ethics"

Usually we think of ethical advice as telling us what to do, or what not to do: Tell the truth. Don't steal. Aristotle's approach is different: it tells us what to *be*, not what to *do*. For Aristotle (and, indeed, for his contemporaries) our actions are seen to flow from our character, and the job of ethical training is to develop our character in such a way that we become "all that we can be" (to steal a phrase from the U.S. Army). Aristotle's approach is often called a "virtue ethics" approach because it focuses on virtues, i.e., traits of good character. I prefer to call it a "be ethic" because it tells us what to be, not what to do. Most ethical theories tell us what to do; they are "do ethics."

The difference between "be ethics" and "do ethics" turns out to be less than it may at first seem. Indeed, for every "do ethic," there is a corresponding "be ethic" and vice versa. Take any proposal for THE GOOD. From that GOOD flows a "do ethic" advising us to act so as to fulfill that good. But so also flows a "be ethic" that advises us to develop our character in such a way as to promote fulfillment of that good.

In the chapters that follow, we could approach each of the central points from a "be ethics" perspective. But "do ethics" is much more familiar and so, for simplicity's sake, that's the approach we will use.

Introduction to *Care, Autonomy, and Justice: Feminism and the Ethic of Care*

Grace Clement

S ince the early 1980s, moral philosophers and social scientists, both feminist and non-feminist, have debated the basis, the normative merits, and the implications of the approach to morality called the ethic of care. The ethic of care emphasizes aspects of moral reasoning that are not generally emphasized by dominant Western moral theories, especially by Kantian ethics. Because these aspects of moral reasoning have been most important in women's traditional activities and experiences, the ethic of care has been of special interest to feminist ethicists. In this work I give an overview of the debate between the ethic of care and the predominant ethic of justice, defend a particular point of view on this debate, and show how this debate and the ethic of care are important for moral and feminist theory. In particular, I argue that the ethic of care is an often neglected but essential dimension of ethics, but that we must make distinctions between versions of the ethic based on their roles in challenging or contributing to women's oppression. Doing so requires that we challenge standard accounts of the relationship between care and justice.

The ethic of care and the ethic of justice are especially worthy of our attention because they are not merely two among many different approaches to ethics. They are more fundamental than other possible ethics because they thematize two basic dimensions of human relationships, dimensions that might be called vertical and horizontal. The ethic of justice focuses on questions of equality and inequality, while the ethic of care focuses on questions of attachment and detachment, and both sets of questions can arise in any context. As Carol Gilligan writes:

> All human relationships, public and private, can be characterized *both* in terms
> of equality and in terms of attachment, and ... both inequality and detachment
> constitute grounds for moral concern. Since everyone is vulnerable both to
> oppression and to abandonment, two moral visions—one of justice and one of
> care—recur in human experience. The moral injunctions, not to act unfairly

toward others, and not to turn away from someone in need, capture these different concerns. (Gilligan 1987, 20)

Traditionally, these two ethics have been kept separate from one another, such that each ethic has focused on one dimension of human relationships to the exclusion of the other. This has resulted in extreme forms of the two ethics, in uncaring forms of justice and unjust forms of care. The fact that these ethics have been gender-coded, reflecting and contributing to relations of dominance and subordination, might lead us to think that they are merely symptoms of these particular social conditions. However, these ethics are not merely reflections of gender, but of fundamental dimensions of human relationships, and thus their relationship to one another is of great importance for morality in general, as well as for questions of gender.

My approach differs from three approaches to the care/justice debate frequently taken: the celebration of the ethic of care as a feminine ethic, the assimilation of the ethic of care to a justice perspective, and the rejection of the ethic of care from a feminist perspective. Proponents of the "feminine" approach have as a general goal the recognition and celebration of women's distinctive activities and experiences. They regard the ethic of care as a creation of women which is usually ignored or devalued by male-defined moral theory. While advocates of this feminine approach do not necessarily believe that all or only women use this ethic, their interest in the ethic arises because (they believe) women especially use it. For instance, Carol Gilligan's work is based on psychological research which she believes demonstrates women's particular use of the ethic of care. Others, like Nel Noddings and Sara Ruddick, do not rely on empirical research, but explore the ethic implicit in and arising out of traditionally female practices like child-rearing.

This feminine approach examines the implications of women's distinctive approach to morality by moving the ethic of care from the periphery to the center of moral theory. Doing so is thought to reveal that the prevailing ethic of justice and its emphasis on autonomy are often dangerous and illusory. The individualism of standard male-defined approaches to morality is replaced by an emphasis on interdependence and the maintenance of relationships. This approach also challenges the traditional understanding of relations between relative strangers in the male-associated public realm as morally paradigmatic, instead focusing on relations between family members and friends in the female-associated personal realm. The abstract universalism of the ethic of justice is replaced by the contextualism of the ethic of care. In short, traditional approaches to ethics tend to dismiss women's distinctive moral orientation. This feminine approach concludes that it is important to give the ethic of care the credit it deserves, in part by showing how it reveals shortcomings in the prevailing ethic of justice.

An obvious question raised by the above approach is: Is the ethic of care really a women's ethic? If we are asked to celebrate the ethic of care because women use this ethic, then we should first be sure that women really use it. A number of critics have argued that no empirical correlation between women and the ethic of care has been demonstrated. They suggest instead that Gilligan and her collaborators "heard" what confirmed the stereotypes of women that they already accepted. Moreover, insofar as Gilligan's "different voice" truly *reflects* anyone's moral orientation, critics argue that it is biased toward the values of the Western, white, well-educated women Gilligan's research has focused on.[1] Similarly, in Noddings's and Ruddick's work, mothers seem to represent all women. Barbara Houston writes:

> The feminist standpoint adopted by Gilligan, Ruddick and Noddings ... appears to assume a form of female essentialism. That is, despite disclaimers by each of them about the dangers of speaking for all women, there does appear to be the assumption that women's experience is similar enough for us to posit a women's ethics arising out of women's distinctive labor. (Houston 1987, 259)

Thus it seems that Gilligan and other advocates of a feminine ethic of care are guilty of the same error Lawrence Kohlberg committed, that of false universalism. While Kohlberg posited the moral experiences of men as *human* moral experience, those defending the ethic of care as a feminine ethic seem to posit the experiences of a specific, nonrepresentative group of women as *women's* moral experience.

This charge of false universalism has linked the debates surrounding the ethic of care to recent feminist debates about the importance of recognizing the differences between women, and about whether despite their differences in race, class, culture, etc., all women share what Marilyn Frye calls "a ghetto of sorts" (Frye 1983, 9). Here, however, I will avoid these debates. I will not be concerned with social-scientific questions about whether women use the ethic of care, and I will not defend the ethic of care on the grounds that women use it. Instead I will focus on questions about the adequacy of the ethic of care as a moral theory. I will ask whether the ethic of care is a satisfactory approach to morality, regardless of who uses it. This does not mean that gender is irrelevant to my study of the ethic of care. Even if many women do not use the ethic of care, this ethic undeniably captures a widely-held view of what women are and ought to be. The ethic of care is socially coded as a *feminine* ethic, while the ethic of justice is socially coded as a *masculine* ethic. We need not make any false generalizations about women to recognize that women's traditional activities and experiences are especially relevant to a study of the ethic of care.

This brings me to the second general approach to the ethic of care that I discuss: The assimilation of the ethic of care to a justice perspective (e.g., Hill 1987, Sher 1987). Moral philosophers who take this approach emphasize the distinction between gender and ethics. They point out that while most historical philosophers have had deplorable things to say about women, this sexism can and should be distinguished from what they have had to say about morality. By restricting our attention to moral questions, they argue, it becomes clear that the debates between "the ethic of care" and "the ethic of

---

1 Patricia Hill Collins does, however, refer to care as Afrocentric (Collins 1990, 215–17).

justice" are merely contemporary versions of familiar moral debates, such as the Kant/Hume debate over the roles of reason and sentiment in morality. Moreover, even if most moral philosophers have had little to say about care issues, their moral theories generally allow for the ethic of care. Even an ethic of justice, like Kant's, which does not focus on care themes can nevertheless encompass them. In fact, a moral theory can be evaluated by the extent to which it can accommodate the ethic of care and/or the moral views of women. As Susan Moller Okin writes:

> The best theorizing about justice has integral to it the notions of care and empathy. ... The best theorizing about justice is not good enough if it does not, or cannot readily be adapted to, include women and their points of view as fully as men and their points of view. (Okin 1989, 15)

According to this approach, those who champion the ethic of care tend to caricature the theories they label ethics of justice, assuming, for instance, that universal principles preclude rather than require a close attention to context. Rather than shifting our focus to a new approach, then, we need to examine ethics of justice more carefully to see whether and how they can accommodate care concerns. According to Kant's moral theory, for instance, autonomy is a fundamental value, but it need not be understood individualistically, or threaten our sense of community. Justice and care should not be understood as alternative approaches to morality, but rather as complementary approaches. This approach argues that justice is the proper ethic for our public interactions, while care is the proper ethic for our interactions with family and friends. In short, according to this "justice" approach, the ethic of care need not be rejected, but neither is it an important development in moral theory. Not only have care themes been emphasized in various historical moral theories, such as those of Aristotle and Hume, but the ethic of care can be assimilated by so-called ethics of justice, such as Kant's moral philosophy.

First, I will briefly respond to the charge that the ethic of care is not significantly different from Aristotelian or Humean ethics. Supporting this charge is the fact that the recent interest in the ethic of care has coincided with a renewed interest in virtue ethics. The fundamental difference between the recent attention to the ethic of care and these other traditions is that study of the ethic of care, at least at its best, has brought critical attention to the gender-coding of our moral concepts. It has clarified and challenged the sexual division of moral labor. Aristotle and Hume also made reference to the gender-coding of moral concepts, but they sought to reinforce rather than challenge the sexual division of moral labor. Contemporary ethicists studying virtue ethics do not do this, but they have for the most part ignored gender issues. At least insofar as the ethic of care has been studied from a feminist perspective, it is a significant departure from Aristotelian, Humean, and contemporary virtue ethics.

The relationship between the ethic of care and the ethic of justice depends on one's characterizations of the two approaches. Some versions of the ethic of care are clearly incompatible with almost any version of the ethic of justice, while other versions of the ethic of care seem compatible with a standard version of the ethic of justice in one way or another. Depending on the precise nature of the two ethics, this complementarity might mean merely that the ethics have distinct spheres of application, or it might mean that the two ethics can be combined into one comprehensive ethic. In this book, I will begin with versions of the ethic of care and the ethic of justice that I will call ideal types. I will focus on three features of the two ethics which are typically emphasized and which serve to define the ethics in opposition to one another. These are the ethics' relative abstractness or concreteness, their priorities, and their conceptions of the self. In particular, I will develop a definition of the ethic of care based on its contextual decisionmaking, its priority of maintaining relationships, and its social conception of the self. In contrast, I will define the ethic of justice in terms of its abstract decision making, its priority of equality, and its individualistic conception of the self. Although I have chosen features that are typical in and which I believe capture the essence of each ethic, my definitions are archetypes that no thinker necessarily holds in the precise forms I have presented. While I will work with versions of the two ethics that are defined in clear opposition to one another, I show that these versions of the ethics are not morally ideal. Instead I will show how the interactions between the ethics can help us sort out better and worse versions of each ethic.

Although I believe that the ethic of justice and the ethic of care are in many ways compatible, I challenge the attempt to assimilate the ethic of care into the ethic of justice. Doing so does not give the ethic of care equal status to the ethic of justice. Instead, it maintains the traditional hierarchy according to which that which is coded as masculine is regarded as more important than that which is coded as feminine. Assimilating care into the ethic of justice *cannot* be done in a way that gives care equal status to justice. It can only be done by interpreting care through the perspective of justice, thereby devaluing and marginalizing it. By maintaining the standard focal points of the ethic of justice, we lose the benefits offered by the focal points of the ethic of care and by the interaction between the ethics' different focal points. Even though the ethic of justice's emphasis on general principles does not preclude attention to context, it creates the impression that general principles are both distinct from and more important than contextual detail. Likewise, while the ethic of justice's individualism does not logically imply that social connections are unimportant, it does have that nonlogical implication (Calhoun 1988, 452).

Thus the care perspective is the central focus of this book. I do not claim that all moral theorists should treat the ethic of care as central. My purpose is to assess the moral value of the ethic of care, and doing so requires that I consider the ethic on its own terms, rather than from the perspective of another approach. While I will work toward integrating the two ethics into a complete account of moral reasoning, I will also remain aware of the real danger that the ethic of care might be assimilated and thus devalued by the ethic of justice.

This brings me to the approach to the ethic of care often taken by feminists. While they acknowledge that the ethic of care is a good ethic in the sense that the world would be a better place if everyone used it, feminists often insist that the important questions do not concern the ethic's intrinsic value, but its social context. In fact, they argue, the ethic of care amounts to a resuscitation of traditional stereotypes of women, stereotypes which are used to rationalize the subordination of women. Joan Williams writes:

> Gender stereotypes were designed to marginalize women. These stereotypes no doubt articulated some values shunted aside by Western culture. But the circumstances of their birth mean they presented a challenge to predominant Western values that was designed to fail, and to marginalize women in the process. (Williams 1991, 97)

According to these critics, the ethic of care is less a creation of women than an unjust demand upon women, as it requires women to take care of men and men's interests at the expense of themselves and their own interests. In other words, the ethic of care compromises the autonomy of the caregiver, and is therefore inconsistent with feminist goals. Moreover, the ethic's restriction to personal contexts means that it is unable to address any large-scale social issues, and thus provides no political resources for challenging women's oppression. In short, according to this approach, the ethic of care is inseparable from women's oppression, and while its celebration may make women feel better about their assigned roles, it still reinforces their subordinate status. As Katha Pollitt writes, "It's a rationale for the status quo, which is why men like it, and a burst of grateful applause, which is why women like it. Men keep the power, but since power is bad, so much the worse for them" (Pollitt 1992, 804).

I think it is important to draw attention to the social context of the ethic of care. But just as it is a mistake to ignore care's social context, it is also a mistake to *reduce* the ethic of care to the distorted ways it is often practiced. We can look for the moral and political possibilities implicit in the ethic of care while actively addressing its dangers. Like those taking the above approach, one of my guiding questions will be: Is the ethic of care helpful or harmful to women? But rather than simply accepting or rejecting the ethic of care, I distinguish between better and worse versions of it. This general approach is not unique to me. Others have asserted that a feminist ethic of care is possible. About the "relational turn" of which the recent interest in the ethic of care is a part, Martha Minow writes:

> Unlike relational thought uninformed by feminist perspectives, feminist work tends to focus also on conflict, power, domination, and oppression as features of relationships. The relational turn thus represents not a denial of or lack of interest in conflict and disunity but a focus on the interpersonal and social contexts in which these and all other human relations occur. (Minow 1991, 198)

Others have also identified necessary conditions for a feminist ethic of care. For instance, Barbara Houston asserts that "If anything is to be declared good, right, or just, it had better be demonstrably good, right, or just for women" (Houston 1987, 261).

My approach expands on such suggestions by focusing on two particular features of the ethic of care. These features are ones which feminists cite as problematic *and* which advocates of the ethic of care consider essential. Because of this conflict, these features serve as fundamental dilemmas for any attempt to develop a feminist ethic of care. The first contested feature is autonomy. As I noted above, feminine advocates of the ethic of care argue that autonomy is an individualistic value that the ethic of care rejects in favor of relational virtues. However, its feminist critics argue that because the ethic of care compromises a caregiver's autonomy, it fails by feminist standards. The second contested feature is the ethic of care's status as a personal ethic, appropriate for our relations with family, friends, or those otherwise close to us, such as students. Again, for its feminine advocates, the ethic's scope gives personal relations the moral attention they deserve, correcting the ethic of justice's view of personal relations as morally insignificant in comparison to public relations. Conversely, its critics argue that a feminist ethic must not be limited to personal relations, and must include a concern for social justice.

I argue that the ethic of care reveals important problems with the concept of autonomy, but that these problems are not present in all versions of autonomy. Likewise, critics are correct to insist on the importance of autonomy, but not all versions of the ethic of care conflict with autonomy. I also argue that advocates of the ethic of care are correct to emphasize the moral centrality of personal relations, but that expanding the boundaries of the ethic of care does not amount to trivializing personal relations. Indeed, it does just the opposite, taking the norms of personal relations as a paradigm for all moral relations. I agree with feminist critics that the ethic of care's personal scope is inadequate, but I argue that the ethic can be expanded beyond this scope in a way that enriches rather than threatens the ethic of justice. In general, then, I argue that the conflicts between care and justice orientations need not lead us to accept one at the expense of the other, indeed, these conflicts can help us distinguish between better and worse versions of each ethic. Most importantly, they allow us to construct a genuinely feminist ethic of care.

Bibliography

Calhoun, Cheshire. 1988. Justice, Care, Gender Bias. *Journal of Philosophy* 85: 451–463.

Collins, Patricia Hill. 1990. *Black Feminist Thought: Knowledge, Consciousness, and the Politics of Empowerment.* Boston: Unwin Hyman.

Frye, Marilyn. 1983. *The Politics of Reality: Essays in Feminist Theory.* Trumansburg, NY: The Crossing Press.

Gilligan, Carol. 1987. Moral Orientation and Moral Development. In *Women and Moral Theory,* ed. Kittay and Meyers, 19–33. Totowa, NJ: Rowman & Littlefield.

Hill, Thomas. 1987. The Importance of Autonomy. In *Women and Moral Theory,* ed. Kittay and Meyers, 129–138. Totowa, NJ: Rowman & Littlefield.

Houston, Barbara. 1987. Rescuing Womanly Virtues. In *Science, Morality and Feminist Theory,* ed. Hanen and Nielsen, 237–262. Calgary: University of Calgary Press.

Minow, Martha. 1991. *Making All the Difference.* Cambridge: Harvard University Press.

Okin, Susan Moller. 1989a. *Justice, Gender, and the Family.* New York: Basic Books.

Okin, Susan Moller. 1989b. Reason and Feeling in Thinking about Justice. *Ethics* 99: 229–249.

Pollitt, Katha. 1992. Are Women Morally Superior to Men? *The Nation:* 12/28/92, 799–807.

Sher, George. 1987. Other Voices, Other Rooms? Women's Psychology and Moral Theory. In *Women and Moral Theory,* ed. Kittay and Meyers, 178–189. Totowa, NJ: Rowman & Littlefield.

Williams, Joan. 1989. Deconstructing Gender. *Michigan Law Review* 87: 797–845. (Also: 1991. In *Feminist Legal Theory,* ed. Bartlett and Kennedy, 95–123. Boulder, CO: Westview Press.)

DISCUSSION QUESTIONS

1. As a patient, by which of these theories would you most hope your doctor would be acting?

2. As a citizen, by which of these theories would you most hope your leaders and public health officials would make decisions?

3. Which of these theories would be most appropriate during a triage situation?

4. Cite two examples in which each theory would be a recommended approach in bioethics.

5. Should we respect a parent's decision for their child, no matter what it is? Why or why not? Specifically, should we allow parents to refuse a lifesaving treatment for their child?

6. How do we know which of these theories are best to use in a situation? If it is because of their results or rules, how do we avoid reasoning that already assumes a consequentialist or deontological structure?

7. Some individuals want to give organs directly to celebrities. Should the celebrity be able to bypass the traditional transplant process to receive a direct donation from someone they have never met?

8. Is it better for us to choose one theory we decide is the best and use it universally? Or is it better to approach each case and examine it uniquely?

Recommended Cases

Dr. Anna Pou—Dr. Pou is a physician who stayed in New Orleans during Hurricane Katrina. During a reverse triage situation, Dr. Pou allegedly euthanized nine patients who would not have been able to survive the storm. This was to reserve treatment for individuals who were more likely to survive.

Belding Scribner and the God Committee—Scribner created the first version of an artificial kidney. The technology was expensive, and there were few of Scribner's artificial kidneys available. A committee was formed to choose among applicants, selecting those who could have the process paid for, were likely to survive, and were the most deserving. Qualifications for "most deserving" included personality, merit, employment, number of children and family members, and the promise of helping others. Based on these qualifications, among others, the committee chose who would and would not receive the lifesaving treatment.

UNIT II

Research

....

Unit II Introduction

The history of medical research can be viewed as a history of progress. This progress, however, must be questioned in a few ways: How was this progress achieved? For whom was this progress? Abstaining from these questions leaves us all at risk. Balancing attempts at progress with human dignity is a necessity. We need not look far to find abuses committed in research; the world has routinely provided us with examples of research without ethics. The history of progress is also a history of failure.

Consider, for instance, the trials conducted by scientists under Germany's Nazi Party in the mid-twentieth century: these included freezing experiments, desalinization studies, and eugenics, among many other horrific trials. In the United States, we ran our own eugenics programs (see Unit V) in addition to the violations experienced by Henrietta Lacks and many individuals like herself. Henry Beecher, in "Ethics and Clinical Research" (1966), recorded countless experiments carried out in the name of research, which included a woman having a melanoma transplanted from her daughter to herself (both women died of metastatic melanoma) and researching whether the central nervous system or cardiovascular system would collapse first by decreasing patients' blood pressures. Stanley Milgram tricked test subjects into thinking they had murdered a man, and student volunteers at Stanford gave us insight to Abu Ghraib through the mental destruction that occurred to students designated as prisoners by students designated as prison guards.

"In Remembrance There Is Prevention: A Brief Review of Four Historical Failures to Protect Human Subjects" records and recalls four experiments in which those upon whom research was done were endangered, particularly the most vulnerable populations. By examining these cases, the need for an ethical standard for research and protection of these individuals can be historically evaluated and pragmatically implemented.

"Ethical Issues when Conducting Research" begins with a history demonstrating the need for ethical research as well as some of the ethical codes, such as the Nuremburg Code and the Belmont Report, which have been used to protect patients against violations. The paper then delves into some of the goals of research as it relates to ethical goals. By following the proposed principles, one can hope to maintain the safety and dignity of patients involved in research.

In "Ethics," we examine our relationships with animals. Beginning with examples of the treatment of chimpanzees, we bridge theory with practice by looking at the traditional approach to animal ethics as well as alternative approaches. "Ethics" concludes by examining the complexities which occur when the social contexts of animal and human interests overlap.

The focus of "Super-Muscly Pigs: Trading Ethics for Efficiency" is on the practice of altering animals for human use. This paper focuses on genetically engineered animals, beginning with pigs that had their genes altered to increase their strength. We are approached with three main concerns: the animals' perspective, the global context, and the changing human-animal relationship.

In Remembrance There Is Prevention

A Brief Review of Four Historical Failures to Protect Human Subjects

Cameron R. Nelson, JD, RN, LLM (cand)

Introduction

Much of the modern economy is impacted by today's increasing globalization, and clinical research is at the forefront. Cost is a significant factor that may propel clinical research beyond U.S. borders (Shah, 2002). It is estimated that the cost per subject for tracking and other administrative requirements is up to ten times more for an American subject as compared with a subject abroad (O'Reilly, 2009). The number of people enrolled in clinical research worldwide is significant; in 2009, that number was 50 million and growing (Shetty, 2009). In the past decade, the percentage of research conducted overseas has continued to increase, and by 2010, more than 50 percent of clinical trial subjects and sites were located overseas (Levinson, 2010). The volume of studies and sometimes-remote locations give rise to the potential for lack of oversight and little recourse for vulnerable populations that may be affected by participation in clinical research.

As the research community and the world reflect on the legacy of the Tuskegee syphilis experiments, it is appropriate, given the global nature of clinical research, to also remember three additional studies in which appropriate protections were not provided to human subjects. From Tuskegee to Nuremberg, and from Guatemala to Nigeria, the research community and the public are reminded of past failures marked by a lack of respect for the human subjects involved. Remembrance, here, is not intended to vilify or to blame, but rather to ensure that adequate human subject protections are never again forgotten, particularly for vulnerable populations.

USPHS Syphilis Experiments in Tuskegee, 1932–1972

During the 1920s, researchers of the U.S. Public Health Service (USPHS) developed plans to study the response of black males to disease, hypothesizing that the response would differ from that of white males. In Macon County, Alabama, nearly forty percent of black males tested positive for syphilis and researchers believed that this community would be ideal to study disease progression.

Although the initial plans included medical treatment, the financial devastation of the Depression in 1929 eliminated project funding. Rather than abandoning all research, USPHS researchers decided to proceed with a limited study (Thomas & Quinn, 1991).

Under the amended research design, USPHS hired Nurse Eunice Rivers to coordinate the study, officially known as "The Tuskegee Study of Untreated Syphilis in the Negro Male." Nurse Rivers, a black woman, was a graduate of the Tuskegee Institute and had worked on various public health projects in Alabama since the early 1920s. By hiring Nurse Rivers, USPHS researchers believed that they could more efficiently reach community members (Thomas & Quinn, 1991).

From the beginning, the study was marked by decisions to withhold information. Men involved in the study were told only that they were being tested for "bad blood," a euphemism for a variety of illness including syphilis, anemia, and fatigue. Each of the 600 men initially involved in the study was tested for syphilis; 399 of these men were found to have syphilis and the remaining 201 men tested negative (Jones, 1993). The men were not informed of either the purpose of the study, or whether they had syphilis; further, no information about modes of transmission or treatment was provided. Researchers intended for the study to conclude after six months of observation; however, the study was extended in an effort to examine the men at regular intervals until their death. By tracking the men through their lives and then performing an autopsy after death, researchers aimed to follow the complete progression of untreated syphilis (Jones, 1993).

Researchers employed several tactics to persuade the men to continue in the study and the families to permit autopsies after the men in the study died. Actions were taken that created an illusion of treatment. For example, the men in the study were given iron, aspirin, and placebo treatments (Jones, 1993). In some cases, this regimen did provide a modest improvement in the men's health and resulted in further study participation. Many of the men involved in the study would have been otherwise unable to afford medical exams and willingly accepted the exams offered in exchange for participation. Researchers also offered free spinal taps, advertising the procedure as a special treatment. While these procedures aided researchers in obtaining information about neuro-syphilis, they were of no therapeutic value to the men involved in the study (Brandt, 1978). In order to encourage families to allow autopsies, family members were approached and offered a small stipend for burial costs in exchange for their permission to conduct an autopsy. As many of the families had limited economic resources for burial expenses, the stipend was a persuasive inducement (Jones, 1993).

Over the 40-year course of the study, treatments for syphilis advanced. At the beginning of the study, the standard treatment of the disease consisted of a combination of arsenic, mercury, and bismuth given over an extended period. By the early 1940s, physicians began treating patients with an abbreviated course, generally one week, of an arsenic derivative and bismuth. Researchers actively coordinated with local clinics to ensure that the men involved in the study did not receive treatment (Jones, 1993). By the mid-1940s, penicillin became the treatment of choice for syphilis, and Alabama law required that anyone testing positive for venereal diseases be treated with appropriate medication

(Jones, 1993). Again, researchers ensured that the men involved in the study were neither tested nor treated as required by law. Researchers also provided the names of men involved in the study to draft boards and physicians treating World War II draftees, requesting that these men not receive the penicillin treatment that was administered to all other draftees diagnosed with syphilis (Thomas & Quinn, 1991). The men involved in the study remained unaware of their diagnosis and the benefit of treatment.

Although the men involved in the study were not given diagnosis and treatment information, the study itself was not kept from the medical community. Articles were published in medical journals and papers on the study were presented at medical conferences, but few, if any, ethical objections were raised to the study protocol and methodology (White, 2000). Over the 40-year span of the study, significant achievements occurred in the protection of human subjects and yet, the study continued without change. A contemporaneous development, the Nuremberg Code, discussed below, set forth guidelines on human research and clearly articulated the requirement for informed consent. During the 1960s and early 1970s, USPHS began developing guidelines for clinical trials and peer review, although the men involved in the study did not appear to benefit from these protections (Jones, 1993).

Some of the first objections to the study arose in 1964, after an article was published describing the 30-year history of the study. A physician wrote to the authors and questioned the ethical approach of the study in denying treatment (Jones, 2000). Despite these questions, no change to the study was made. Five years later, a USPHS employee raised additional ethical concerns, and USPHS convened a review panel, which consisted of white physicians, all but one of whom were already familiar with the study's existence and methods. The only panel member who recommended physical examinations and treatment for the men involved in the study was the physician who was previously unaware of the study (Jones, 1993). Further compounding the panel's disturbing analysis was a determination that the socioeconomic status and education level of the men involved in the study rendered informed consent impossible. The panelists subsequently decided to ask the Macon County Medical Society for consent, substituting the judgment of the medical board members for that of the men involved in the study. Finally, in 1972, a reporter for the Associated Press, Jean Heller, reported the details of the study. The article resulted in outrage and ultimately led to study termination, after 40 years of experiments involving human subjects who neither consented, nor were given appropriate protections (White, 2000).

As with other human subject studies conducted without appropriate protections, the impact of the USPHS syphilis experiments in Tuskegee did not end when the study itself concluded. Perhaps one of the most significant pejorative results from the study is distrust of the medical community, research community, and the U.S. Government, particularly among minorities. While some steps have been taken to restore trust, ranging from regulations to apologies, it is clear that there must be continued vigilance to overcome this legacy of the study. Ultimately, this study, as with the other studies discussed in this article, serves as a powerful reminder of what occurs when researchers fail to recognize that each human subject is worthy of respect.

The Nazi Holocaust Experiments and the Resulting Dcotors' Trial at Nuremberg, 1946–1947

Following World War II, twenty-three Nazi physicians and administrators were accused of organizing and participating in war crimes and crimes against humanity by way of medical experiments and procedures to which prisoners and civilians were subjected unnecessarily, and prosecuted between 1946 and 1947. At the conclusion of the trial, seven defendants were convicted and later executed, nine defendants were convicted and sentenced to terms of incarceration, and seven defendants were acquitted. As a result of the Doctors' Trial, the Nuremberg Code was produced and served as one of the first significant human rights documents ("Nuremberg Code," 1949).

The defendants in the Doctors' Trial were indicted on four specific counts: conspiracy to commit war crimes against humanity, war crimes, crimes against humanity, and membership in a criminal organization; the medical experiments exposed during trial detailed the horrific nature of the research conducted primarily on Nazi concentration camp inmates ("Final Report," 1949). The experiments varied in duration, beginning in 1939 and concluding in 1945. The defendants argued that these experiments were necessary for the German defense, and therefore justifiable; however, this argument did not prevail at trial. The experiments included: (1) sterilization experiments, to determine large scale methods that would ensure the elimination of specified populations and maximize prisoners' work; (2) high-altitude experiments, conducted for the German air force and designed to intentionally cause subject death; (3) freezing experiments, also conducted for the German air force to examine treatments for hypothermia and also designed to cause subject death; (4) seawater experiments, conducted for the German navy and air force to test methods of making seawater safe to drink; (5) poison experiments, conducted to evaluate the effect of poisons in food and bullets; (6) incendiary bomb experiments, conducted to test treatments for phosphorus burns that involved intentionally inflicting burns prior to treatment; (7) jaundice experiments, conducted for the German armed forces to examine causes and inoculations against jaundice; (8) bone, muscle, nerve, and bone transplant experiments, conducted to benefit German armed forces; (9) mustard gas experiments, conducted for the benefit of the German armed forces, in which physicians deliberately inflicted wounds on prisoners and then infected them with mustard gas; (10) Sulfanilamide experiments, conducted for the benefit of the German armed forces to test drug effectiveness after wounds were deliberately inflicted to resemble battlefield wounds; (11) vaccine experiments, conducted for the benefit of German armed forces to test vaccine efficacy for typhus, smallpox, malaria, and other diseases; (12) tubercular Polish national imprisonment and homicide, conducted under the pretext of protecting German health; (13) skeleton collection for anatomical research; (14) euthanasia, beginning at German asylums and continuing to the Nazi concentration camps. The prosecution delivered its opening statement on December 9, 1946, and on June 2, 1948, the seven defendants sentenced to death were executed ("Final Report," 1949).

The judges of the Doctors' Trial rejected the notion that these experiments were necessary and acceptable, even in a wartime setting. In their condemnation of the experiments, the court in the Doctors' Trial articulated what we now know as the Nuremberg Code. The Code specifies ten critical points to govern human subject research. The precise words of the Nuremberg Code serve as powerful reminders, especially as the reader reflects on other studies discussed herein.

1. The voluntary consent of the human subject is absolutely essential. This means that the person involved should have legal capacity to give consent; should be so situated as to be able to exercise free power of choice, without the intervention of any element of force, fraud, deceit, duress, over-reaching, or other ulterior form of constraint or coercion; and should have sufficient knowledge and comprehension of the elements of the subject matter involved as to enable him to make an understanding and enlightened decision. This latter element requires that before the acceptance of an affirmative decision by the experimental subject there should be made known to him the nature, duration, and purpose of the experiment; the method and means by which it is to be conducted; all inconveniences and hazards reasonable to be expected; and the effects upon his health or person which may possibly come from his participation in the experiment.

The duty and responsibility for ascertaining the quality of the consent rests upon each individual who initiates, directs or engages in the experiment. It is a personal duty and responsibility which may not be delegated to another with impunity.

2. The experiment should be such as to yield fruitful results for the good of society, unprocurable by other methods or means of study, and not random and unnecessary in nature.

3. The experiment should be so designed and based on the results of animal experimentation and a knowledge of the natural history of the disease or other problem under study that the anticipated results will justify the performance of the experiment.

4. The experiment should be so conducted as to avoid all unnecessary physical and mental suffering and injury.

5. No experiment should be conducted where there is an a priori reason to believe that death or disabling injury will occur; except, perhaps, in those experiments where the experimental physicians also serve as subjects.

6. The degree of risk to be taken should never exceed that determined by the humanitarian importance of the problem to be solved by the experiment.

7. Proper preparations should be made and adequate facilities provided to protect the experimental subject against even remote possibilities of injury, disability, or death.

8. The experiment should be conducted only by scientifically qualified persons. The highest degree of skill and care should be required through all stages of the experiment of those who conduct or engage in the experiment.

9. During the course of the experiment the human subject should be at liberty to bring the experiment to an end if he has reached the physical or mental state where continuation of the experiment seems to him to be impossible.

10. During the course of the experiment the scientist in charge must be prepared to terminate the experiment at any state, if he has probable cause to believe, in the exercise of the good faith, superior skill, and careful judgment required of him that a continuation of the experiment is likely to result in injury, disability, or death to the experimental subject ("Nuremberg Code," 1949).

It is striking to note that the timing of the Doctors' Trial and subsequent Nuremberg Code intersected both the sexually transmitted disease study in Guatemala and the Tuskegee syphilis study. Despite the fact that the articulation of the Nuremberg Code did not immediately change the course of the aforementioned studies, it is a marker of the birth of American bioethics and has shaped the development of human research protections in the ensuing decades (Annas, 2009).

USPHS Sexually Transmitted Diseases Inoculation Study, Guatemala, 1946–1948

While researching the Tuskegee syphilis studies, Susan M. Reverby, a historian at Wellesley College, uncovered documents detailing the U.S. Public Health Service (USPHS) sexually transmitted diseases inoculation studies that were conducted in Guatemala between 1946 and 1948 (Reverby, 2011). In an attempt to determine whether taking penicillin after sexual intercourse would prevent the spread of syphilis, gonorrhea, and chancroid (Stein, 2010), researchers were funded by a grant from the National Institutes of Health, then a part of USPHS ("Discovery of 1940s Study," 2010). In order to explore reliable methods of infecting subjects, researchers utilized three different subject populations—male prisoners, mental patients, and soldiers—and introduced the subjects to prostitutes already infected with one of three sexually transmitted diseases, or intentionally infected by researchers ("Ethically Impossible," 2011). Researchers also attempted to develop inoculations derived from *Treponema pallidum*, the bacteria that causes syphilis infection. In total, 128 subjects were infected with chancroid, 234 subjects were infected with gonorrhea, and 427 subjects were infected with syphilis. Researchers also drew blood samples from 438 children for future research (Stein, 2010).

Among the facts of the Guatemala studies that strike the conscience is that the research occurred during the same time as the Doctors Trial of the Nuremberg Tribunal, discussed above. The Tribunal issued its opinion in August 1947, in the middle of the Guatemala studies, and the Nuremberg Code was finalized in 1949, only one year after the Guatemala studies concluded. Despite the contemporaneous trial at Nuremberg, human subjects research was governed by few regulations that would have benefitted the Guatemalan subjects. Although the Guatemala study notably lacked an IRB review, no requirement for such a review existed at that time; indeed, IRBs were not codified until nearly 30 years later ("Federal Policy").

The Guatemala study used human subjects from three populations defined today as vulnerable: prisoners, the mentally ill, and wards of the state ("Historian Who Unveiled," 2010). Much like the lack of IRB requirement, though, these populations did not benefit from modern protections during the study. Prisoners were not defined as a vulnerable population until 30 years after research concluded in Guatemala. The risks to which prisoners were subjected were magnified by their lack of freedom and, in many cases, left them more inclined to agree to activities with higher risk, such as receiving inoculations and engaging in sexual intercourse with prostitutes. Casting even more doubt on the prisoners' level of willing participation in the study is the fact that researchers offered rewards, in the form of cigarettes, to prisoners who participated ("Ethically Impossible," 2011). Just as prisoners participating in human subject research received protections much later, so, too, did wards of the state and the mentally ill. For both populations, federal regulations were promulgated 30 or more years after the Guatemala study.

Unlike IRBs and vulnerable population protections, informed consent was the ethical norm during the time of the Guatemala study and yet, researchers infected the subjects with sexually transmitted diseases without their knowledge and without obtaining informed consent ("Ethically Impossible," 2011). Researchers not only failed to inform subjects that the inoculations they received contained the syphilis bacteria, but actively deceived subjects and institutional officials by leading them to believe that the inoculations were actually drugs with medicinal benefit. Instead of obtaining informed consent from the research subjects, researchers preferred to work with the Guatemalan government and institutional officials where the subjects were housed ("Ethically Impossible," 2011). In order to gain cooperation and permission from the institutions, researchers promised to provide medications, medical treatments, and other desired commodities, such as refrigerators. Researchers demonstrated shocking indifference to the suffering of subjects. In one case, a woman who was infected with syphilis had reached a terminal state as a result of the disease. Rather than providing treatment, researchers poured gonorrhea-infected pus into her eyes and other orifices, and also attempted to infect her again with syphilis. She later died ("Ethically Impossible," 2010). Several documents expose the reservations that USPHS doctors had about the study, namely that the mentally ill could not give consent and that there were moral concerns about their conduct; however, the Guatemala study continued (Reverby, 2011).

After two years of experiments, the Guatemala studies concluded, 71 subjects had died, and researchers had made little progress to answer their threshold question (Stein, 2010). In November 2010, President Barack Obama charged the Presidential Commission for the Study of Bioethical Issues (PCSBI) to further investigate the facts of the Guatemala study and to conduct a thorough review of human subjects protection. The Commission's report in September 2011 found that the Guatemala study involved unconscionable violations of ethics, even when judged against the requirements of the medical ethics of the time of the study. PCSBI concluded that researchers intentionally exposed over 1,000 people to sexually transmitted diseases without obtaining appropriate consent prior to the commencement of research. A particularly disturbing fact to PCSBI and to the public was the knowledge that many of the researchers in Guatemala had participated in earlier studies involving prisoners in Indiana. In the earlier studies, the researchers did obtain consent from the subjects, but chose to proceed without consent in Guatemala ("Ethically Impossible," 2011). The researchers' actions demonstrated an unacceptable double standard for consent with foreign national research subjects and highlighted the need for adequate safeguards.

Pfizer Trovan Study, Nigeria, 1996

Nearly fifty years after researchers left Guatemala, the actions of one of the largest American pharmaceutical companies once again highlighted the paucity of safeguards for human subjects abroad, particularly for vulnerable individuals. In 1996, Nigeria experienced an epidemic of bacterial meningitis (Abdullahi v. Pfizer, 2009), an infection of the membranes of the brain and spinal cord. Left untreated, bacterial meningitis carries a mortality rate of nearly ten percent. In an effort to gain FDA approval for pediatric use of its new antibiotic, trovafloxacin mesylate or Trovan, Pfizer sponsored a study in Kano, Nigeria (Abdullahi v. Pfizer, 2009). Approximately 200 children were recruited to serve as subjects for the research protocol testing Trovan (Abdullahi v. Pfizer, 2009). The fallout from the flawed study and the violations of research ethics ultimately led Pfizer to withdraw its FDA application for use of Trovan as a treatment for bacterial meningitis in pediatric patients (Stephens, 2006).

From its inception, the Pfizer study appeared rushed and indifferent to human research safeguards. Despite an average research protocol development timeframe of one year, the Pfizer protocol was completed in six weeks, even though this study was the first in which Trovan, in oral form, was tested in children (Stephens, 2000). This rush to protocol completion left several research requirements unmet, notably, a formal IRB or ethics committee approval was never obtained for the study (Khan, 2008; Stephens, 2006). In failing to obtain IRB approval, researchers not only violated statutory requirements, but also eliminated an opportunity to design a study that provided appropriate measures to prevent similar side effects as were reported from testing Trovan on animal subjects (Abdullahi v. Pfizer, 2009). During Trovan animal testing, significant side effects were reported, including liver damage, degenerative bone conditions, joint damage, and cartilage damage (Stephens, 2000). Several critics have commented that IRB review of the Pfizer protocol would likely have led to significant modifications to study design. Rather than submit to the IRB review and approval process, the principal investigator forged a letter, backdated to the week before the experiment began, stating that the study had been pre-approved by the hospital's ethics committee (Stephens, 2006; Wollensack, 2007). It was later discovered that not only was the review omitted, but that an ethics committee did not exist at the hospital on the date of the forged letter (Stephens, 2006). Although Pfizer claimed that the principal investigator, a Nigerian physician, led the study, an independent Nigerian investigative committee found that the physician served only nominally in this role; rather, the committee found that researchers based in the United States directed the study (Stephens, 2007). After confessing to the forged ethics committee approval letter, the physician also claimed that he was unaware of the experiment's results and did not see any research data until the investigative committee showed him study results (Abdullahi v. Pfizer, 2009).

Beyond the failure to obtain IRB approval, the Pfizer protocol exposed the pediatric subjects to unnecessary and preventable risks. Researchers allegedly failed to conduct preliminary testing to determine whether the subjects had a definitive diagnosis of meningitis and whether the strain of meningitis affecting Kano, Nigeria, was responsive to Trovan. Further, researchers allegedly did not screen pediatric subjects for joint and liver problems and exclude them from the study, despite the results of Trovan animal testing, nor did they transfer subjects who failed to respond to Trovan to the control group (Abdullahi v. Pfizer, 2009). The protocol itself called for lumbar punctures, also known as spinal taps, and blood tests to assess the efficacy of Trovan; however, these tests were not completed (Stephens, 2006). Claims have also been made that researchers took steps to render the control group treatment ineffective by reducing the dosage of ceftriaxone, the antibiotic received by the control group, thereby elevating the comparative efficacy of Trovan; Pfizer denies intentionally using a dose lower than recommended for the control group (Stephens, 2007).

Adding to the flaws in Pfizer's Trovan study are allegations that its researchers failed to obtain appropriate informed consent in accordance with statutory requirements and ethical guidelines (Abdullahi v. Pfizer, 2009). The pediatric subjects and their guardians were allegedly unaware that they were participating in research (Stephens, 2000). The native language of Kano is Hausa, and although Pfizer claims that Nigerian nurses at the hospital where the subjects were treated explained the study details in the appropriate language, the nurses did not fully translate the consent form (Stephens,

2000). Pfizer was unable to produce any signed consent forms for the Trovan study; further, Pfizer admitted that no witnesses signed documentation attesting that verbal consent was given (Stephens, 2000; "Statement of Defence," 2007). It has also been alleged that the subjects' guardians often asked the nurses or researchers to make a decision for them because they did not understand the conversation (Shah, 2002). As part of the consent process, Pfizer also had a duty to disclose appropriate alternative treatment; however, it is unclear whether researchers informed patients of the free, conventional, and reportedly effective treatment that was concurrently provided by Doctors Without Borders in the same hospital where the Pfizer study occurred ("Statement of Defence," 2007).

Pfizer's trouble with consent did not stop with the subjects and their guardians. In its attempt to seek FDA approval for testing, Pfizer represented that it had obtained appropriate approval of the local Nigerian government and the hospital's ethics committee, as previously discussed ("Statement of Defence," 2007). The Nigerian investigative committee later found that Pfizer did not obtain approval from the Nigerian government to administer Trovan and accordingly, the study was an "illegal trial of an unregistered drug" which violated Nigerian law (Stephens, 2006). Although Pfizer claims that it relied on approval letters from the Food and Drug Administration and Control (NAFDAC), the Nigerian FDA counterpart, NAFDAC's director reported that the agency was unaware of the Pfizer study (Stephens, 2006).

At the conclusion of the Pfizer Trovan study, which lasted only a few weeks, 11 children had died and others suffered symptoms attributable to meningitis, including blindness, deafness, seizures, immobility, and disorientation (Stephens, 2000). No definitive causal link between Trovan and the deaths that occurred has been determined, although many are quick to point out that a few of the children died a few hours or days after taking Trovan (Stephens, 2000). In at least one case, a pediatric subject's symptoms worsened during the study, and yet she continued to receive the drug for three additional days, ultimately dying while still receiving Trovan (Stephens, 2006). Although Pfizer firmly states that any deaths were the result of meningitis infection, not the treatment provided, the Nigerian investigative committee found that the research protocol deviations compromised the care of the pediatric research subjects (Stephens, 2006; "Statement of Defence," 2007). In June 1997, the FDA inspected Pfizer's files and cited the company for record-keeping deficiencies ("Trovan's Troubled History," 2000). Four months later, in October 1997, Pfizer withdrew its FDA application for the use of Trovan as a treatment for bacterial meningitis in pediatric patients ("Trovan's Troubled History," 2000).

Conclusion

In their own right, each of the studies reviewed in this article is a powerful reminder of the tragedy that occurs when research does not deem every human subject worthy of the most basic respect. Examining the failures of these studies to provide appropriate and adequate protections for human subjects reminds the modern research community to move forward, considering these lessons. Collectively, these studies are marked by failure to obtain consent, failure to educate, and failure to respect. Research abroad has become more common and research administrators must bear in mind factors that create potential for harm, such as lack of understanding due to language barriers, socioeconomic differences that may entice subject participation, to factors yet unknown. Failures in human subject protection and respect, whether occurring in foreign or domestic research, impact community perception of the research community's integrity. Remembering the lessons of these four tragic historical events serves as a powerful aid in prevention.

References

Abdullahi v. Pfizer, Inc., 562 F.3d 163, 169 (2d Cir. 2009).

Annas, G. J. (2009). The legacy of the Nuremberg Doctors' Trial to American bioethics and human rights. *Minnesota Journal of Law, Science, & Technology, 10*(1), 19–40.

Benedek, T. G. & Erlen, J (1999). The scientific environment of the Tuskegee study of syphilis, 1920–1960. *Perspectives in Biology and Medicine, 43*(1), 1–30.

Brandt, A. M. (1978). Racism and research: the case of the Tuskegee syphilis study. *Hastings Center Report December*, 21–29.

Brawley, O. W. (1998). The study of untreated syphilis in the Negro male. *International Journal of Radiation Oncology Biology Physics, 40*(1), 5–8.

Brody, B. A. (1998). *The ethics of biomedical research: an international perspective.* New York: Oxford University Press.

Caplan, A., & Annas, G. J. (1999, July 2). Tuskegee as metaphor. *Science, 285*(5424), 47.

Corbie-Smith, G. (1999). The continuing legacy of the Tuskegee syphilis study: considerations for clinical investigation. *American Journal of Medical Science, 317*(1), 5–8.

Discovery of 1940s study ethics breach leads to apology from NIH, investigation (2010, October 6). *Medical Research Law & Policy Report.* Retrieved from http://news.bna.com/mrln/display/batch_print_display.adp

Epstein, A. M. & Ayanian, J. Z. (2001). Racial disparities in medical care. *New England Journal of Medicine, 344*(19), 1471–1473.

Ethically impossible: STD research in Guatemala from 1946 to 1948 (2011). Presidential Commission for the Study of Bioethical Issues. Retrieved from http://www.bioethics.gov/cms/sites/default/files Ethically-Impossible_PCSBI.pdf.

Fairchild, A. L. & Bayer, R. (1999). Uses and abuses of Tuskegee. *Science, 284*(5416), 919–921.

Federal policy for the protection of human subjects ('Common Rule'). *U.S. Department of Health and Human Services.* Retrieved from http://www.hhs.gov/ohrp/humansubjects/commonrule/index.html

Final report to the Secretary of the Army on the Nuremberg war crimes trials under control council no. 107 (1949).

Freimuth, V. S., Quinn, S. C., Thomas, S. B., Cole, G., Zook, E., & Duncan, T. (2001). African Americans' views on research and the Tuskegee syphilis study. *Social Science and Medicine, 52,* 797–808.

Grodin, M. A. (1993). Medicine and human rights: a proposal for international action. *Hastings Center Report, 8,* 8–9.

Harvard Law School Library, *Nuremberg trials project digital document collection, introduction to NMT case 1, U.S.A. vs. Karl Brandt, et al.* Retrieved from http://nuremberg.law.harvard.edu/php/docs_swi.php?DI=1&text=medical

Historian who unveiled 1940s STD studies on Guatemalan prisoners recounts findings (2010, November 3). *Medical Research Law & Policy Report.* Retrieved from http://news.bna.com/mrln/display/batch_print_display.adp

Jones, J. H. (1993). *Bad blood: the Tuskegee syphilis experiment.* New York: The Free Press.

Jones, K. (2006, May 18). A flawed approach to drug testing. Washington Post, p. A22. Retrieved from http://www.washingtonpost.com/wp-dyn/content/article/2006/05/17/AR2006051701902.html

Kahn, J. P., Mastroianni, A. C., & Sugarman, J. (1998). Beyond consent: seeking justice in research. New York: Oxford University Press.

Khan, F. (2008). The human factor: Globalizing ethical standards in drug trials through market exclusion. *DePaul Law Review, 57,* 877, 899.

Kampmeier, R. H. (1974). Final report on the "Tuskegee Syphilis Study." *Southern Medical Journal, 67*(11), 1349–1353.

Levinson, D. R. (2010). Challenges to FDA's ability to monitor and inspect foreign clinical trials. *Office of the Inspector General at the Department of Health and Human Services*, June 2010, ii. Retrieved from http://oig.hhs.gov/oei/reports/oei-01-08-00510.pdf

Nuremberg Code (1949). *Trials of War Criminals Before the Nuremberg Military Tribunals Under Control Council Law*, 10, 82. Retrieved from http://www.ushmm.org/research/doctors/Nuremberg_Code.htm

Postal, S. W. & Diaz, R. W. (2011). After Guatemala and Nigeria: The future of international clinical research regulation. *The Health Lawyer, 24*(1), 1, 7–20.

O'Reilly, K.B. (2009). Outsourcing clinical trials: Is it ethical to take drug studies abroad? *American Medical News*, Sept. 7, 2009. Retrieved from http://www.ama-assn.org/amednews/2009/09/07/prsa0907.htm

Reverby, S. M. (2011). "Normal exposure" and inoculation syphilis: A PHS "Tuskegee" doctor in Guatemala, 1946–1948. *Journal of Policy History, 23*(1), pp. 6–28.

Reverby, S. M., Ed. (2000). *Tuskegee's truths: rethinking the Tuskegee syphilis study.* Chapel Hill: University of North Carolina Press.

Shah, S. (2002, July 1). Globalizing clinical research. *The Nation, 26.*

Shetty, P. (2009, July 11). With all of their ethical obstacles, are clinical trials really worth it? [Review of the book *Chasing medical miracles*, by A. O'Meara]. *New Scientist, 48.*

Shuster, E. (1997). Fifty years later: the significance of the Nuremberg Code. *New England Journal of Medicine, 337,* 1436–1437.

Statement of Defence, Attorney General of Kano State v. Pfizer International Incorporated et al., High Court of Kano State, Suite No. K/233/2007. Retrieved from http://www.pfizer.com/files/news/trovan_litigation_statement_defense.pdf

Stein, R. (2010, October 1). U.S. apologizes for newly revealed syphilis experiments done in Guatemala. *Washington Post*. Retrieved from http://www.washingtonpost.com/wp-dyn/content/article/2010/10/01/AR2010100104457.html?sid=ST2010100104522

Stephens, J. (2000, December 17). Where profits and lives hang in balance. *Washington Post*, p. A01. Retrieved from http://www.washingtonpost.com/wp-dyn/content/article/2007/0702/AR2007070201255.html

Stephens, J. (2006, May 6). Panel faults Pfizer in '96 clinical trial in Nigeria. *Washington Post*, p. A01. Retrieved from http://www.washingtonpost.com/wp-dyn/content/article/2006/05/06/AR2006050601338.html

Stephens, J. (2007, June 2). Pfizer faces new charges over Nigerian drug test. *Washington Post*, p. D01. Retrieved from http://www.washingtonpost.com/wp-dyn/content/article/2007/06/01/AR2007060102197.html

Thomas, S., & Curran, J. W. (1999). Tuskegee: from science to conspiracy to metaphor. *American Journal of the Medical Sciences, 317*(1), 1–4.

Thomas, S. B., & Quinn, S. C. (1991). The Tuskegee syphilis study, 1932–1972: implications for HIV education and AIDS risk education programs in the black community. *American Journal of Public Health, 81*(11), 1498–1505.

Trials of war criminals before the Nuremberg Military Tribunals under Control Council Law (1949). 2(10), 181–182.

Trovan's troubled history: Early development (2000, December 17). *Washington Post*. Retrieved from http://www.washingtonpost.com/wp-dyn/content/article/2008/10/01/AR2008100101376.html

White, R. (2000). Unraveling the Tuskegee study of untreated syphilis. *Archives of Internal Medicine, 160*, 585–598.

Wollensack, A. F. (2007). Closing the constant garden: The regulation and responsibility of U.S. pharmaceutical companies doing research on human subjects in developing nations. *Washington University Global Studies Law Review, 6*, 747, 756.

21 C.F.R. § 50.52(c) (2011).

45 C.F.R. § 46.306(a)(2)(ii) (2011).

45 C.F.R. § 46.306(a)(2)(iii) (2011).

45 C.F.R. § 46.306(a)(2)(iv) (2011).

45 C.F.R. § 46.409(a)(1) (2011).

45 C.F.R. § 46.409(a)(2) (2011).

45 C.F.R. § 46.409(b) (2011).

48 Fed. Reg. 9, 814 (Mar. 8, 1983).

Ethical Issues When Conducting Research

Deborah Zelizer, Kathleen McGoldrick, and Deborah Firestone

....

Chapter Goals:

The Goals of This Chapter Are:

- To provide working definitions of the terms *beneficence, justice, respect for persons*, and *informed consent*.
- To explore the relationship between early research practices and the creation of research standards in response to these early research practices.
- To furnish students with an understanding of the Institutional Review Board process and its implications for ethical research.

Learning Objectives:

By the Completion of This Chapter, the Student Will Be Able To:

- Define the terms: *assent, autonomous, Belmont Report, beneficence, coercion, ethical research, informed consent, Institutional Review Board, justice, respect for persons, undue influence, and vulnerable*;
- Explain the process and elements of informed consent;
- Identify ethical issues in clinical research trials that led to the implementation of Institutional Review Boards; and
- Apply the information learned regarding informed consent and the Institutional Review Board to identify ethical issues in case studies.

Introduction

This chapter discusses the role of ethics in research and the creation of standards to ensure that research involving human participants is ethical.

What is ethics in research, and why is it important? In order to answer this question, it's important to provide a summary of some of the early studies conducted under the guise of research, which then served as an impetus for the development of standards for conducting ethical research with human participants. Figure 2.1 illustrates the time line of unethical research practices that led to current research standards in the United States.

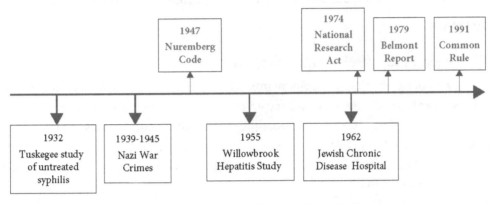

Figure 2.1 The time line of events leading to current research standards.

In terms of the research process, the highest level of ethical behavior is mandatory during each of the five stages of research. The primary focus of this chapter will be to explain the policies and practices that a researcher engages in to ensure that research participants (also known as human subjects) are protected during the design and empirical stages of research (Table 2.1).

Table 2.1 Stages of Research: Action Steps Highlighted in This Chapter

Stages	Action Steps
Design (Planning)	Selecting the best research design and research methods (sampling, data collection, data analysis) that align with research question(s), **submitting the research study proposal for IRB review**
Empirical (Doing)	**Obtaining IRB approval to conduct the study**, recruiting/selecting the sample, **obtaining informed consent from participants, collecting the data**

Ethical Research—Defined

Ethical behavior, as defined by Merriam-Webster's (2017) dictionary, is "conforming to accepted standards of conduct." Research, as defined by the Office for Human Research Protections (OHRP) in Chapter 1, is "a systematic investigation, including research development, testing and evaluation, designed to develop or contribute to generalizable knowledge" (OHRP, 2008, p. 2). *Ethical research*, therefore, is a systematic investigation designed to develop or contribute to generalizable knowledge that conforms to accepted standards of conduct.

Early Research Practices and the Creation of Research Standards

The Tuskegee Study of Untreated Syphilis

In 1932, the Tuskegee Institute, in conjunction with the U.S. Public Health Service, began the *Tuskegee Study of Untreated Syphilis* (also known as the Tuskegee Syphilis Study or Tuskegee Syphilis Experiment) to document the natural disease progression of syphilis in black men. Six hundred black men from Macon County, Alabama, most of whom were living in poverty, were enrolled in the study. Of these 600 men, 399 had syphilis; 201 did not (Centers for Disease Control and Prevention [CDC], 2013; Rothman, 1982).

Men who participated in the study were not told they were in an experiment; they were told they were being treated for "bad blood," which for some involved painful spinal taps. Participants were given "free medical exams and treatment," meals, and, upon death, a burial stipend was paid to their survivors. In 1945, penicillin was approved by the U.S. Public Health Service (USPHS) to treat syphilis, but the Tuskegee Study continued without the men being treated (CDC, 2013; Rothman, 1982).

Rothman (1982) explains that the study continued "through the 1960s, untouched by the civil rights agitation, and unaffected by the code of research ethics adopted by the USPHS itself. It ended only in 1972, when an account of the experiment in the Washington Star sparked a furor" (p. 5).

Nazi War Crimes

Between 1939 and 1945, medical researchers in Nazi Germany conducted a wide range of heinous and often fatal experiments on prisoners of concentration camps, without their consent. Other prisoners were euthanized, solely because they were deemed "unworthy of life" (Nazi Medical Experiments, n.d.).

In 1946, an American military court held criminal proceedings against 23 German physicians for their participation in war crimes and crimes against humanity. These proceedings, known as the Nuremberg Doctors' Trial, lasted 140 days. Eighty-five witnesses testified, and 1,471 documents were introduced. Although 16 of the physicians were found guilty, they could not be found guilty of violating the rights of human participants because, at that time, there were no accepted standards (Nazi Medical Experiments, n.d.; Shuster, 1997).

The Nuremberg Code

The *Nuremberg Code* was framed by American judges sitting in judgment of the Nazi doctors accused of committing heinous medical experiments in concentration camps. This code combined Hippocratic ethics and the protection of human participants into a single document and has been called the most important document in the history of ethics in medical research. The code focuses on the human rights of research participants and is comprised of ten standards that physician-investigators must adhere to when conducting experiments on human participants. Included in this code was the requirement of voluntary informed consent; the recognition that risks must be weighed against anticipated benefits; and the caveat that experiments should only be conducted by scientifically qualified persons (Fischer, 2006; Shuster, 1997).

Willowbrook Hepatitis Study

The **Willowbrook Hepatitis Study** was initiated in the mid-1950s at the Willowbrook State School, an institution for cognitively disabled children on Staten Island, New York. Hepatitis was widespread at the institution, and researchers were interested in studying the natural course of viral hepatitis as well as the efficacy of gamma globulin (Krugman, 1986).

Although parents gave consent for their children to be inoculated with a mild form of hepatitis, they were not fully informed of the possible hazards involved in the study. There is evidence that indicates parents might have been under the impression that if they did not give consent, their children would not be cared for. Researchers defended their actions under the guise that because hepatitis was widespread at the institution, the children most likely would have been infected with the disease within their first year at the institution (Krugman, 1986; Rothman, 1982).

Jewish Chronic Disease Hospital Study

Another example of an egregious study that took place in an institution in which people were to have been cared for is the **Jewish Chronic Disease Hospital Study**. In 1963 two doctors injected live cancer cells into hospitalized patients with chronic diseases. The premise of the study was to see if patients who were debilitated with a chronic disease rejected cancer cells, albeit at a slower rate, than healthy patients (McNeil, 1993, p. 57).

It is important to note that this study was so egregious that it was brought to the attention of the Board of Regents of the State University of New York who, upon review, found that not only had the research protocol not been presented to the hospital's review committee, but the patients' physicians were unaware of their patients' involvement in this study. As a result of their blatant disregard for the welfare of the patients who, unknowingly, had been injected with live cancer cells, the doctors were found guilty of "fraud, deceit and unprofessional conduct in the practice of medicine" (Mulford, 1967, p. 99).

National Research Act

The development of the regulatory process governing the ethical conduct of researchers began with the signing of the **National Research Act** into law in 1974, thereby creating the National Commission for the Protection of Human Subjects of Biomedical and Behavioral Research. The commission was charged with identifying the key components of ethical research involving human participants and developing guidelines to ensure that human research is conducted in accordance with those principles (OHRP, 1979).

Belmont Report

The **Belmont Report** was drafted in 1979 by the National Commission for the Protection of Human Subjects of Biomedical and Behavioral Research. The Belmont Report is important to the content of this chapter as it identified three important basic principles: **respect for persons**, **beneficence**, and **justice**, to be followed in the ethical conduct of research on humans.

Respect for Persons

"Respect for persons incorporates at least two ethical convictions: first, individuals should be treated as autonomous agents, and second, that persons with diminished autonomy are entitled to protection" (OHRP, 1979, para. 3). According to the Belmont Report, Part B: Basic Ethical Principles, "An autonomous person is an individual capable of deliberation and personal goals and of acting under the direction of such deliberation" (OHRP, 1979, para. 3). Demonstrating respect for the decisions made by an **autonomous person** would involve respecting their decisions and opinions, unless said decisions and opinions would be harmful to others. Showing a lack of respect to an autonomous person could be manifested in a number of ways. One could show a lack of respect by rejecting, or interfering with, a person's ability to carry out or act on their opinions and choices. Another example would be by withholding information, for no compelling reason, for the purpose of interfering with an individual's ability to make a decision. Two hallmarks of an autonomous individual would be that they have the ability to both understand and process information and, should they choose to participate in a research study, they are free to do so without being coerced or influenced by others. "In research involving human subjects, respect for persons implies that, when given adequate information about the research project, that subjects voluntarily decide to participate" (OHRP, 1979, para. 3).

Not every person has the capacity to act as an autonomous agent, whether due to illness, a mental disability, or circumstances that severely restrict their freedom and therefore may require protection while incapacitated. The term **diminished autonomy** is used to describe an individual who is not able to act as an autonomous agent and therefore is not "... capable of deliberation and personal goals and of acting under the direction of such deliberation" (OHRP, 1979, para. 3). This diminished autonomy may be manifested as having limitations when it comes to giving thoughtful consideration to or carrying out their personal goals.

A **vulnerable population** can be described as "the disadvantaged sub-segment of the community ..." (Shivayogi, 2013, p. 53). When a person has limitations on either their capacity or voluntariness, they are considered **vulnerable**. "The vulnerable individuals' freedom and capacity to protect one-self from intended or inherent risks is variably abbreviated, from decreased free will to inability to make informed choices" (Shivayogi, 2013, p. 53).

Examples of participants who lack capacity or are unable to make their own choices and decisions are children, those with intellectual disability, prisoners, students in hierarchical organizations, institutionalized individuals, the elderly, and individuals who are educationally and economically disadvantaged (Shivayogi, 2013).

Coercion

Under Part C, section 1 of the Belmont Report, *coercion* "occurs when an overt threat of harm is intentionally presented by one person to another in order to obtain compliance" (OHRP, 1979, para. 11). Consider the following example: an elderly woman who is a resident in a nursing home is forced to choose between participating in a research study or leaving the nursing home. The elderly woman lacks the ability to make a decision based on her own free will. She is being forced to choose one of two options; participate in the research study and stay in the nursing home or don't participate in the research study and leave the nursing home. The participant in this case is being threatened in order to obtain compliance, the threat that she will not be able to stay in the nursing home. Her ability to make a decision based on her own free will has been taken away.

Another example of coercion in research would be where a physician threatens to stop providing care to his patient unless the patient joins a clinical trial. The physician is making an overt threat of harm: "You can no longer be my patient" in order to coerce or force the patient to participate in the study.

Undue Influence

In contrast, Part C, section 1 of the Belmont Report defines **undue influence** as influence that "occurs through an offer of an excessive, unwarranted, inappropriate or improper reward or other overture in order to obtain compliance" (OHRP, 1979, para. 10). Consider the example of an investigator promising students in her psychology class that they will receive extra credit if they participate in her research project. If students are presented with only this one way to earn extra credit, then the investigator is unduly influencing potential study participants. If, however, students who did not want to participate in the research project were given nonresearch opportunities to earn extra credit, then the possibility of undue influence is decreased.

Another example of undue influence would be if a researcher offered a large sum of money (for instance, a month's salary) to participants for taking part in a one-day study to test the effects of a drug with potentially serious side effects that is under investigation. Because the sum of money offered could persuade potential participants to engage in the study against their better judgment, this offer could present undue influence.

Beneficence

The principle of beneficence requires that persons are treated in an ethical manner by (1) protecting them from harm; and (2) maximizing possible benefits and minimizing possible risks of harm. It is the obligation of researchers to maximize benefits for the individual participant and/or society while minimizing the risk of harm to the individual participant. This doesn't mean that there are not any risks involved to participants. It means that thoughtful consideration has been given to both the possible benefits and harms, and a decision is then made as to when it is justifiable to seek certain benefits in spite of the risks involved and when the risks involved outweigh the potential benefits. Should the risks outweigh the benefits, consideration should be given to determining if there is another way to conduct a study in which the same knowledge could be obtained with lower risks to participants (OHRP, 1979).

Justice

The principle of justice in Part B raises the question: "Who ought to receive the benefits of research and bear its burdens?" (OHRP, 1979, para. 11). An injustice occurs when a person who is entitled to a benefit is denied said benefit, without good reason, or when a burden is unduly imposed (OHRP, 1979). The selection of research participants must be fair, avoiding participants from a population (i.e., educationally or economically disadvantaged) or selecting participants from a certain population only for the experiment group. Research conducted in the United States in the early-to-mid-20th century illustrates the violation of the principle of justice. For example, participants in the Tuskegee Study of Untreated Syphilis were disadvantaged rural black men who were denied treatment, even though penicillin was available to treat syphilis, so that the study could be continued.

Common Rule

Using the Belmont Report as an ethical guideline, currently what governs the protection of human subjects in the United States is the Federal Policy for the Protection of Human Subjects, also known as the **Common Rule**. Said another way, the Common Rule operationalized the principles in the Belmont Report by setting the rules and procedures researchers must follow when conducting human subject research. The rules and procedures include, but are not limited to, developing an Institutional Review Board (IRB), setting standards on what information must be included in the consent forms, and the level of review studies must go through. There is more information on this in later sections of the chapter. This 1991 federal policy requires compliance across 15 different federal departments and agencies. Each department/agency was required to develop a set of policies that complied with this federal regulation (i.e., Department of Education, 34 CFR Part 97; Department of Justice, 28 CFR Part 46; National Science Foundation, 45 CFR Part 690; etc.). The policies for human subjects' protection in health science research can be found under the Department of Health and Human Services regulations, 45 CFR Part 46. This regulation has four subparts which include: "subpart A, also known as the Federal Policy or the 'Common Rule'; subpart B, additional protections for pregnant women, human fetuses, and neonates; subpart C, additional protections for prisoners; and subpart D, additional protections for children" (OHRP, 2016, para 2).

Proposed updates to the Common Rule are being worked on; it is anticipated that these changes will go into effect in 2019. Some of the proposed changes being discussed include revisions to the consent forms that will help people be better informed when making decisions about whether to participate in a research study as well as inclusion of initiatives to help streamline the review processes for low-risk research studies and the paperwork required annually to renew low-risk research studies (OHRP, 2018).

References

Centers for Disease ~~trol~~ and Prevention. (2013, December). *The Tuskegee timeline*. In U.S. Public Health Service ~~Tuskegee. Retrieved from https://www.cdc.gov/tuskegee/time-line.htm

Fisc~~ ary of important documents in the field of research ethics. *Schizophrenia Bulletin,* ~~hbul/sbj005~~

~~rook hepatitis studies revisited: Ethical aspects. *Clinical Infectious Diseases,*~~

Me~~ *thical behavior*. Retrieved from https://www.merriam-webster.com/

McNe~~ *ethics and politics of human experimentation*. New York, NY: Cambridge University Press~~

Mulford, R. D. (1967). Experimentation on human beings. *Stanford Law Review, 20*(1), 99–100.

Nazi Medical Experiments. (n.d.). *Holocaust Encyclopedia*. Retrieved from https://www.ushmm.org/wlc/en/article.php?ModuleId=10005168

Office for Human Research Protections. (1979, April 18). *The Belmont Report: Ethical principles and guidelines for the protection of human subjects of research*. Retrieved from https://www.hhs.gov/ohrp/regulations-and-policy/belmont-report/index.html

Office for Human Research Protections. (2008). *Engagement of institutions in human research*. Retrieved from https://www.hhs.gov/ohrp/regulations-and-policy/guidance/guidance-on-engagement-of-institutions/index.html

Office for Human Research Protections. (2016, March 18). *Federal policy for the protection of human subjects ("Common Rule")*. Retrieved from https://www.hhs.gov/ohrp/regulations-and-policy/regulations/common-rule/index.html

Office for Human Research Protections. (2018, June 18). *HHS and 16 Other Federal Departments and Agencies Issue a Final Rule to Delay for an Additional 6 Months the General Compliance Date of Revisions to the Common Rule While Allowing the Use of Three Burden-Reducing Provisions during the Delay Period.* Retrieved from https://www.hhs.gov/ohrp/final-rule-delaying-general-compliance-revised-common-rule.html

Rothman, D. J. (1982). Were Tuskegee & Willowbrook "studies in nature"? *Hastings Center Report, 12*(2), 5–7. doi:10.2307/3561798

Shivayogi, P. (2013). Vulnerable population and methods for their safeguard. *Perspectives in Clinical Research, 4*(1), 53–57.

Shuster, E. (1997). Fifty years later: The significance of the Nuremberg Code. *New England Journal of Medicine, 337*(20), 1436–1440. doi:10.1056/NEJM199711133372006

ETHICS

Alice Crary

....

Anyone who sets out to grapple with questions about animals and ethics is likely to feel pressure to justify themselves. When children are starving, journalists are beheaded, and young women are kidnapped and forced into marriage, it can seem frivolous to take an interest in how animals are treated. There are long-standing ethical traditions that may seem to support this dismissive attitude—traditions that represent animals as morally indifferent objects, that is, as things that do not in themselves call for respect and whose flesh is therefore available to be cultivated, caged, cut, measured, prodded, maimed, dismembered, displaced, displayed, harvested, and devoured in any manner as long as it suits recognized human purposes (Regan and Singer 1989). Today ethical approaches that in this way deny animals moral standing are increasingly on the defensive. Yet they still have outspoken defenders (e.g., Oderberg 2000), and the image they bequeath to us of animals as mere disposables that don't in themselves call for solicitude remains operative on a massive scale in settings such as confined feeding operations, industrial slaughterhouses, aqua-farms, hunting grounds on land and in the oceans, zoos, laboratories, and sites of the large-scale conversion of tropical and other forests.

Struck by the need to challenge not only the handling of animals in these settings but also the bodies of thought that seem to support it, people who are concerned with the mistreatment of animals frequently start by trying to show simply that animals do in fact impose on us direct claims for particular forms of treatment and so do have moral standing. The same thinkers then sometimes go on to discuss how, in the nonideal messiness of actual cases, animals should be treated. Part of what makes many real-world cases messy has to do with the fact that human-animal interactions often occur in contexts in which human beings are themselves subject to serious and sometimes intersecting forms of bias (e.g., racist and classist bias). These contexts may well confront us with difficult questions about the possibility of unsentimental and clear-sighted concern for animals that does not neglect the plight of human beings. One strategy for managing these kinds of complexities is to bracket them at the outset and focus initially on what is involved in simply bringing out that animals have moral standing. Different approaches to this basic task encode different assumptions about what ethical thought about animals is like as well as about the kinds of methods proper to it, and it is helpful to start by asking what view of these matters we should adopt.

Two Examples from the Lives of Chimpanzees

Consider, to begin with, the following anecdote that moral philosopher and activist Lori Gruen tells in a recent paper:

> One of the chimpanzees I know living in sanctuary is now over forty years old. Every time humans come around he makes a ridiculous facial expression—he pops both of his lips out and folds them back so the inside of his lips show, making him look clown-like, with a big pink mouth. This chimpanzee was used in the entertainment business and presumably making himself look absurd garnered laughs and attention. He was undoubtedly taught to do this when he was young either by rewarding him when he did or, more likely, punishing him when he didn't. (Gruen 2014, 231)

Suppose that we are inclined to say that Gruen's chimp acquaintance was wronged by the treatment he received. This by itself is already a nontrivial claim, since it presupposes that the chimp is a being with moral standing who merits specific forms of treatment and who can be mistreated. But it still leaves unanswered many questions about the nature of the wrong or wrongs at issue. Was the chimp wronged by being subjected to the pain of punishment-based training? Was he also wronged by being held in captivity? Does the fact that he was forced to make a spectacle of himself for human beings—and that he was left in a condition in which he compulsively reenacts the spectacle—represent, as Gruen suggests, a further and distinct wrong? How significant are these wrongs? And, lastly, what resources do we have in ethics for answering these types of questions?

When animal advocates take a stand on what justifies us in representing animals such as this chimp as having moral standing, they are at the same time at least tacitly addressing this last question. We can see this by turning to what might aptly be called *traditional approaches to animal ethics*. This label can be used to collectively pick out a family of views about animals' moral standing that, while in many respects divergent, resemble each other in accepting a metaphysical picture (i.e., a picture of what kinds of things there are) that is ingrained in our intellectual culture and that provides the framework within which most research in ethics is now pursued. The picture is one in which the empirical or observable world is as such devoid of moral values. Ethical positions that encode this picture have arresting implications for how we conceive the resources available to us for moral thought. Here there is no question of our needing the exercise of moral capacities such as moral imagination in order to get aspects of the empirical world in view in a manner relevant to ethics. That is, there is no question of our needing to look at things from alternative cultural or historical perspectives or to imaginatively enter the experience of individuals very different from us in order to see some things clearly. The image of the empirical world with which we operate in ethics is supposed to be something that, far from being produced through this kind of moral effort, is handed down to us from disciplines such as the natural sciences, where these disciplines are conceived as independent of ethics. Moreover, since the overall image of the empirical world with which we operate in ethics necessarily subsumes within it any image that we have of the worldly lives of animals, this means that the task of bringing the worldly lives of animals into focus in ethics turns out to be a task that, instead of belonging to ethics proper, gets outsourced to disciplines outside it.

Acceptance of this outlook has a notable effect on the way in which we approach ethical questions that come up in reference to Gruen's example of the chimp. It obliges us to regard the project of arriving at the kind of empirical understanding of this creature's life that is relevant to ethics as one that we can entrust to disciplines such as biology, where these are conceived as external to ethics. Now we have to say that there is no room for an engagement with morally saturated accounts of chimpanzee existence to contribute necessarily to this project. Among the things that we thus exclude as essentially irrelevant is engagement with the many charged accounts that the primatologist Jane Goodall has given of her encounters with individual chimps. In one essay, Goodall writes about Jojo, an adult male chimpanzee who was born in the African forest and who, when she met him, had spent ten years in a medical research laboratory in a five-foot-by-five-foot steel cage. Goodall describes what human beings have done to Jojo and other captive chimps in these terms:

> [We have] deprived them of freedom, stole from them the dim greens and browns; the soft gray light of that African forest, the peace of afternoon; when the sun flecks through the canopy and small creatures rustle and flit and creep among the leaves. Deprived them of the freedom to choose, each day, how they would spend their time, and where and with whom. Deprived them of the sounds of nature, the gurgling of streams, murmuring wind in the branches, of chimpanzee calls that ring out so clear, and rise up through the tree tops and drift away in the hills. Deprived them of their comforts, the soft leafy floor of the forest, the springy, leafy branches from which sleeping nests can be made. (Goodall 2000, xii)

Goodall is here trying to impress on us the magnitude of the harm done to Jojo and other chimpanzees in captivity by giving us an evocative account of the humble glory of chimp life in the forest. It would be possible to allow the kinds of responses invited by Goodall's words to shape our efforts to bring the worldly existence of Gruen's chimp into focus. But if we accept an outlook in which the world is as such bereft of moral value, we will be compelled to reject at the outset any nonneutral perspectives that Goodall's writing invites us to adopt as incapable of contributing internally to the empirical understanding of Gruen's chimp that we want in ethics.

Traditional Approaches to Animals and Ethics

This outlook shows up within many of the most well-known and widely discussed approaches to animal ethics in the form of the idea that the fabric of the world is free of moral values. This idea is, for instance, at play in Peter Singer's *Animal Liberation* (Singer 1975). Singer's strategy for showing that animals matter, here and elsewhere, through to the present day, starts from a form of utilitarianism, a view on which the right action in a particular context is the one that best promotes the interests of all creatures concerned, where a creature's "interests" are an expression of her capacity for pain or pleasure. Singer's work has had a major impact, providing inspiration for many subsequent interventions in animal ethics, and what has been most influential is not his utilitarian stance itself but one of its core presuppositions. It is a presupposition of Singer's stance that any consideration a—human or nonhuman—creature merits is a reflection not of its membership in any group (say, her membership in the group "human beings" or in the group "horses") but rather of her individual capacities (e.g., her individual capacities for pain and pleasure). This presupposition provides the unifying principle for the most prominent family of approaches to animal ethics. Many animal advocates follow Singer's lead in treating moral status as a function of individual capacities while at the same time disagreeing with him about which individual capacities are morally relevant (e.g., suggesting that the capacity not for pain but for, say, subjecthood is the mark of moral standing; Regan 1983). Further, although it is in principle possible to endorse a doctrine that grounds moral status in individual capacities without assuming that the world is morally neutral, animal advocates who incline toward such doctrines in fact overwhelmingly integrate an assumption along these lines into their outlooks. Some—notably, Singer—even make their attachment to the assumption explicit (Crary 2016). The result is that members of this set of thinkers treat the identification of the individual worldly capacities that, as they see it, are morally significant as a job that is properly assigned not to ethics itself but to disciplines beyond it. Their preferred approaches to animal ethics thus clearly count as "traditional" in the above sense.

Advocates of these traditional approaches to animal ethics are sometimes referred to collectively *as moral individualists* (Rachels 1990; McMahan 2005). What speaks for this terminology is the fact that these thinkers ground an animal's—nonhuman and human—moral standing in a certain kind of attention to her as an individual (specifically, in morally neutral attention to her individual capacities of mind). However, as will emerge below, there are other thinkers who take up questions of animals and ethics and who, while differing substantially from Singer et al. in their views, nevertheless resemble them in basing their conclusions about moral standing on attention to individual creatures. In light of this convergence, it makes sense to withhold the generic label "moral individualism" from the projects of Singer and like-minded others and to place their work instead under the heading of *traditional moral individualism.*

It is not hard to see how Singer and others use the tenets of traditional moral individualism to show that animals are proper objects of moral concern. Once we're equipped with the thought that moral standing is a function of neutrally available individual capacities, we need do little more than allow that any capacities we regard as morally relevant in human beings are similarly morally relevant in animals. To be sure, this position only seems to have direct implications for the treatment of animals if it is combined with the thought that some human beings and some animals are in fact equally well endowed with morally relevant capacities. Taking their cue from this observation, animal advocates who favor traditional moral individualism often take the further step of drawing attention to cases of human beings who (as a result of illness, injury, age, or some congenital condition) are severely cognitively disabled and who are no better equipped than some animals with what the thinkers in question see as "morally significant capacities." Sometimes these thinkers suggest, for instance, that a person who has suffered a serious brain injury may be no better mentally endowed than a dog or a pig. With reference to the—morally objectionable—idea that severely cognitively disabled individuals are "marginal" cases of humanity, this argument is sometimes referred to as the *argument from marginal cases* (Dombrowski 1997). Wanting to avoid this offensive terminology, some theorists now speak of the *argument from species overlap* (Horta 2014). But a powerful case can be made for thinking that without regard to whether it is *described* in morally problematic terms, this argument *is* morally problematic—indeed outrageous—for implying that severely cognitively impaired individuals merit diminished respect by virtue of their impairments (Kittay 2010; 2016). Setting aside this topic, the point is that this argument is employed by members of the high-profile group of animal protectionists who appeal to the principles of traditional moral individualism to support their view that some animals merit moral attention.

It is a premise of the so-called argument from species overlap that the mere fact of being human (i.e., without regard to the level of one's individual capacities) is morally insignificant. Animal advocates who run versions of the argument, and who thereby effectively contest the tendency to treat being human as by itself morally important, sometimes attack this tendency under the heading of *speciesism* (Ryder 1989). Speciesism is typically understood as unwarranted prejudice in favor of one's species. What traditional moral individualists add to this is the thought that we express such unwarranted prejudice if we suggest that the mere fact of being human matters morally. It would, however, be wrong to assume that champions of all traditional approaches to animal ethics regard as speciesist the idea that simply being human is morally important.

One notable traditional approach that proceeds along different lines is developed in the work of Christine Korsgaard, a moral philosopher who draws her main inspiration from Kant's moral theory. Kant's stated view of animals is that they are mere moral instruments, that is, things that cannot themselves be harmed and that are only subject to "harms" that are indirect reflections of wrongs to human beings (Kant 1996, 192; 1997, 212–13). Korsgaard sets out to show that, despite this expressed view of animals, Kant actually equips us to show that animals matter. She stresses that she thinks a substantial part of the appeal of Kant's core claims about ethics has to do with the fact that they are consistent with the "traditional" idea that the world is a place with no trace in it of moral value or, as she puts it, a "hard" place (Korsgaard 1996b, 4). One of Korsgaard's explicit goals is to mount a modified Kantian defense of animals' moral standing that is likewise consistent with this idea. Her strategy centers on an adaptation of Kant's approach to establishing the moral standing of human beings. She tells us, in recognizably Kantian style, that when we act there is no way for us to avoid placing value on the natures we have as beings capable of rational action and that we are therefore obliged, on pain of inconsistency, to respect the rational natures of all rational beings. She goes on to echo Kant in representing all human beings (and not only those individuals actually endowed with reason) as having rational natures. Her aim in making this further Kantian move is to affirm that all human beings, and not merely the mentally well endowed, have moral standing and that the plain fact of being human is morally important (see chap. 20). Admittedly, there is significant disagreement about whether it is possible, within the framework of Kant's moral theory, to successfully argue for the view that bare humanity is morally important (Kain 2009; McMahan 2008), and Korsgaard herself appears to have advocated different argumentative strategies at different times (Korsgaard 2004, 1996a). Without entering into the dispute about the success of particular attempts to defend in Kantian terms the idea that merely being human matters, it is fair to say that Korsgaard's position patently departs from that of traditional moral individualists who reject this idea as "speciesist."

It does not, however, follow that Korsgaard's enterprise is anti-animal. On the contrary, Korsgaard claims that a subtle adjustment of her Kantian argument for establishing the moral standing of human beings serves to establish the moral standing of animals. Her suggestion is that when we act, there is no way for us to avoid valuing our own animal natures and that we are thus obliged, on pain of inconsistency, to respect the animal natures of all animate beings. That is how Korsgaard makes the case that animals matter, and, as she herself stresses, she makes it without assuming that moral values are part of the furniture of the universe (Korsgaard 2004). Her work thus presents another clear case of a traditional approach to animal ethics.

At this point, it should be clear that the set of traditional approaches to animal ethics comprises different strategies for dealing with the sorts of questions about animals and ethics that we confront when we consider, say, the treatment of Gruen's chimp acquaintance. Whereas traditional moral individualisms invite us to approach ethical questions about the chimp's treatment by asking whether we are granting him the same consideration that we extend to those individual humans who are equivalently endowed with whichever capacities we take to be morally significant, Korsgaardian Kantianisms invite us to proceed by asking whether we are placing the same value on his animal nature that, in acting, we invariably place on our own animal natures. These different strategies may, in some cases, lead to different moral conclusions about the treatment of animals such as Gruen's chimp. Whereas traditional moral individualism, such as Singer's, that ground moral standing in the capacity for pain don't have room for the idea of harms to a creature that the creature himself does not register, and whereas these views thus encourage us to repudiate the idea—championed by

Gruen—that forcing this chimp to make a spectacle of himself in front of humans represents a distinct injury, beyond the injuries he suffered in being caged and subjected to painful training methods, Korsgaardian Kantianism may well have room for this idea.

Alternative Approaches to Animals and Ethics

Despite differences that exist among the ways in which traditional approaches call on us to address ethical questions about the treatment of animals, striking similarities remain. These approaches agree that the empirical understanding of animals' lives that we are after in ethics is the business of disciplines independent of ethics and that moral capacities cannot play a necessary role in this understanding. This is worth underlining because the operative image of what is involved in bringing animals empirically into view in ethics, while dominant, has outspoken critics. There are contributors to animal ethics who directly contest the image specifically by defending the idea that the exercise of moral capacities such as moral imagination is internal to our ability in ethics to arrive at an undistorted view of the world and, by the same token, internal to our ability to arrive at an undistorted view of the worldly lives of animals. Insofar as they thus represent the exploration of ethically charged perspectives as capable of immediately informing our efforts in ethics to get a clear view of aspects of reality, the champions of these oppositional projects adopt a stance that is also characteristic of many Marxist, feminist, and black theories of knowledge or epistemologies, intellectual projects that are sometimes referred to collectively as *alternative epistemologies* (Mills 1998). What distinguishes the animal advocates who sound these themes—and who might aptly be described as fans of *alternative approaches to animal ethics*—from other alternative epistemologists is the conviction that the features of reality that are ethically charged, and that are such that moral effort is required to illuminate them, include, in addition to aspects of rational human social life, also aspects of not wholly rational, animate (i.e., animal as well as human) life.

The debate between traditional and alternative approaches to animal ethics reaches back to the early years of the contemporary animal protection movement—back to the 1970s—and represents an intellectual divide within the movement that is easily one of the most fundamental and one of the most fruitful to explore. Among the noteworthy pioneering alternative interventions in this debate are Mary Midgley's (1983) *Animals and Why They Matter* and some early papers of Cora Diamond's (Diamond 1991, chaps. 13, 14). Within Diamond's writings an alternative approach to animal ethics is especially clearly worked out.

Diamond is explicit about rejecting an understanding of the world in which moral concepts are "hard," that is, "given for, or given prior to, moral thought and life" (Diamond 2010, 56), and she thus repudiates the philosophical outlook distinctive of traditional approaches to animal ethics. She presents us with a view of the world of moral concern that is instead "cloudy and shifting," and on which we need the exercise of moral capacities such as moral imagination to bring it into focus. The features of the world that she thinks require moral exertion to get in view include both human beings and animals, and she attempts to show that they "are not given for [ethical] thought independently of ... a mass of ways of thinking about and responding to them" (Diamond 1991, 327). Elaborating on this point, she argues that ethical thought about human beings and animals of different kinds is shaped by concepts of them that, far from being merely biological, are imaginative and ethically loaded, and she tells us that when in ethics we attempt to capture the worldly circumstances of an individual creature, such a concept is invariably internal to our attempts. This means that, as Diamond sees it, human beings and animals figure in moral thought as beings who merit respect and attention simply as the kinds of creatures they are. So for Diamond—and in this respect she is representative of fans of alternative approaches to animal ethics—there is no question of opposing as "speciesist" the idea that merely being human (or merely being an animal of some kind) matters.

This is not to say that there are no noteworthy differences between, on the one hand, Diamond and other champions of alternative approaches and, on the other, those champions of traditional approaches, such as Korsgaard, who similarly treat the plain fact of being human as morally important. For Diamond and other fans of alternative approaches, human beings and animals only show up for us as mattering insofar as we look at their worldly lives through ethical conceptions, that is, insofar as we use methods that Korsgaard and other champions of traditional approaches are committed to rejecting as distorting. One corollary of this alternative posture is that a mode of instruction that gives new shape to our imaginative sense of the lives of animals of a given kind may directly contribute to the kind of—genuine—understanding of those animals that we seek in ethics. Diamond's work is peppered with examples of bits of writing that help to make a kind of direct contribution of our understanding of human and animal existence. For example, she discusses a poem of Walter de la Mare's that encourages us to look on a titmouse not as a merely biological thing but as "a tiny son of life," and she tries to bring out how, by virtue of thus shaping our attitudes, the poem may be immediately relevant to efforts to arrive at the sort of empirical understanding that is the business of ethics (Diamond 1991, 473–74).

The last decades have witnessed the emergence of a significant variety of projects that, like Diamond's, count as alternative approaches to animal ethics. This includes, among others, projects that derive their inspiration in part from critical theory (Sanbonmatsu 2011), critical race theory (Kim 2015), and, perhaps most conspicuously, feminist theory and the ethics of care (Donovan and Adams 2007; Gruen 2015). The common thread running through these projects is the idea that bringing the worldly lives of animals into focus in ethics requires a type of moral exertion that involves not only arriving at a conception of the things that matter in the lives of the animals in question, but also surveying those animals through the lens of that conception. Brought to bear on the case of Gruen's chimp acquaintance, the idea is that if we are to be able to clearly perceive his worldly circumstances in a manner relevant to ethics, we need to look at him through the lens of an ethical conception of chimp life, say, the sort of conception we might arrive at by engaging with works, like some of Jane

Goodall's, that aim to impart a sense of what flourishing chimp life is like. It's worth pausing to note that the point is not that every attempt to give us a conception of what matters in the lives of chimpanzees will contribute internally to our ability to bring them empirically into focus in ethics. On the contrary, any particular attempt to foster such an ethical conception (such as, say, an attempt by Jane Goodall) may turn out to be manipulative or sentimental or distorting in some way. Still, the act of critically exploring novel conceptions of what matters in chimp life is taken to be integral to legitimate, world-guided moral thought about chimps. For the idea is that we need to have reference to a reasonably sophisticated conception if, for instance, we are to be able to determine whether some treatment to which the chimp has been subjected is a painful prod rather than a playful interaction or whether a given allocation to him of space and resources represents a continuation of captivity rather than sanctuary.

Insofar as we do in fact look at Gruen's chimp acquaintance in the relevant ethically inflected way, he will show up for us as meriting certain forms of treatment simply by virtue of being a chimpanzee. Notice that here our conclusions about the moral standing of the chimp are licensed by a certain kind of—morally nonneutral—attention to him as an individual. So it would not be unreasonable to talk in this connection about a *nontraditional* or *alternative moral individualism*. Abstracting from the terminological question of whether we should in fact speak of "alternative moral individualisms," it is also worth noticing that views that call for the morally nonneutral modes of attention that are in question encounter no problem—of the sort encountered by traditional moral individualists such as Singer who ground moral standing in neutrally accessible individual capacities—about allowing that the chimp may have suffered injuries that he himself can't register. So there is no obstacle to saying that the chimp was wronged by being forced to pop his lips out in a ridiculous grimace and to thereby make a spectacle of himself in front of human beings.

Advocates of traditional approaches to animal ethics have not devoted much energy to responding to alternative approaches, for the most part simply dismissing them as lacking in argumentative rigor. Alternative approaches call on us to look at the worldly lives of animals from specific ethical perspectives and, to the extent that advocates of traditional approaches comment on the work of their alternative counterparts, they tend to charge them with thereby recommending distorting modes of thought. This is because advocates of traditional approaches assume that the empirical world is as such devoid of moral values, and, granted this assumption, ethically nonneutral perspectives appear to have a necessary tendency to block our view of how things are. But it doesn't follow that advocates of traditional approaches have somehow succeeded in formulating a satisfactory rejoinder to advocates of alternative approaches. The dismissive attitude of advocates of traditional approaches seems question begging given that alternative thinkers reject the very metaphysical assumption that is supposed to justify their approach, viz, the assumption that the world is morally neutral. Advocates of alternative approaches reject this assumption and lay claim to a contrasting view of reality as including moral values, and they also hold that, in opposition to what their traditional correlates maintain, the worldly texture of human and animal lives is suffused with such values. So it appears to them that we have to use methods that are morally loaded if we are to arrive at the sort of empirical understanding of animal life that is the business of ethics. The upshot is that there is in fact a substantive quarrel between advocates of traditional and alternative approaches to animal ethics, specifically, a quarrel that turns on questions about how to conceive ethically the worldly lives of animals and the demands of knowing them.

Further Complexities

Discussions about these topics are yet more complex than talk of a broad opposition between "traditional" and "alternative" approaches might seem to suggest. Alongside traditional approaches to animal ethics, which tell us that we are limited to neutral methods in our attempts to bring empirical animal existence into focus, and alternative approaches, which counter that we require morally nonneutral methods, there are also—to mention one further prominent set of views—*poststructuralist* approaches. These approaches resemble alternative approaches insofar as they claim that any empirical methods we use in ethics are invariably morally nonneutral but also differ from them in giving this claim a skeptical inflection, suggesting that it follows that none of our world-directed modes of thought can lay claim to objective authority (Derrida 2008; Haraway 2003; Wolfe 2003). What thus emerges is that there are further wrinkles and subtleties to conversations about how in ethics to construe the challenges of knowing the worldly lives of animals. Nevertheless, for all their intricacy and abstractness, there is a sense in which these conversations are of interest to anyone concerned with the ethical treatment of animals not least because we invariably position ourselves within them when we grapple with questions about how an animal such as, for example, Gruen's chimp acquaintance should be regarded and, where relevant, treated. In grappling with such questions, we cannot help but commit ourselves to a view about the kinds of resources that are available to us for thinking morally about animals.

The view we arrive at will have implications for how we assess social contexts in which human as well as animal interests are at play and in which wrongs to animals are structurally connected with wrongs to human beings. We encounter such complexity when, for instance, we consider the treatment of animals within the practices of members of socially subjected human groups, say, the treatment of gray whales in revived versions of the traditional hunt of the Native American Makah (Kim 2015) or the treatment of animals of various kinds in the sacrificial rituals of Florida-based practitioners of the Santeria religion (Casal 2003). We also encounter such complexity when we set out to think about the treatment of animals in society-wide, modern institutions such as the industrial slaughterhouse, where in the United States since roughly the late 1990s the people who are on "kill floors," where the actually killing and butchering is done, and who are likeliest to be accused if there are allegations of cruelty to animals, are disproportionately men who are either Hispanic immigrants or African

Americans who have been driven by need and lack of better opportunities to take nonunionized jobs that are dangerous, physically demanding, and poorly remunerated (Schlosser 2002; Compa 2005; and Pachirat 2011). How should we approach issues having to do with the treatment of animals that arise in these sorts of complex cases? If we proceed in the style of traditional approaches to animal ethics, then we will see the task of arriving at an adequate empirical understanding of the relevant circumstances as one that falls outside the purview of ethics, and we will take it that the core job for ethical thought is to deploy whatever ethical theory we favor. If, instead, we proceed in the style of alternative approaches, we will see the task of arriving at an adequate empirical understanding of the relevant circumstances as one that requires nonneutral methods (such as, e.g., methods that involve carefully capturing and critically exploring any relevant historical and cultural perspectives), and we will take it that this task is a core project for ethical thought. But without regard to whether we choose a version of one of these two broad approaches—or whether instead we choose a version of some hybrid or wholly different approach—the approach we select will affect not only the way in which we proceed but also the moral conclusions we draw about the treatment that animals of different kinds merit and about the kinds of harms they can be—and are—made to suffer.

Suggestions for Further Reading

Cavell, Stanley, Cora Diamond, John McDowell, Ian Hacking, and Cary Wolfe. 2008. *Philosophy and Animal Life*. New York: Columbia University Press.

Coetzee, J. M. 1999. *The Lives of Animals*. Edited by Amy Gutmann. Princeton, NJ: Princeton University Press.

Gruen, Lori. 2011. *Ethics and Animals: An Introduction*. Cambridge: Cambridge University Press.

Hearne, Vicki. 2007. *Adam's Task: Calling Animals by Name*. New York: Skyhorse.

References

Casal, Paula. 2003. "Is Multiculturalism Bad for Animals?" *Journal of Political Philosophy* 11 (1): 1–22.

Compa, Lance. 2005. "Blood, Sweat, and Fear: Workers' Rights in U.S. Meat and Poultry Plants." Human Rights Watch, January 24. https://www.hrw.org/report/2005/01/24/blood-sweat-and-fear/workers-rights-us-meat-and-poultry-plants.

Crary, Alice. 2016. *Inside Ethics: On the Demands of Moral Thought*. Cambridge, MA: Harvard University Press.

Derrida, Jacques. 2008. *The Animal That Therefore I Am*. Edited by Mary-Louise Mallet and translated by David Wills. New York: Fordham University Press.

Diamond, Cora. 1991. *The Realistic Spirit: Wittgenstein, Philosophy, and the Mind*. Cambridge, MA: MIT Press.

———. 2010. "Murdoch the Explorer." *Philosophical Topics* 38 (1): 51–85.

Dombrowski, Daniel. 1997. *Babies and Beasts: The Argument from Marginal Cases.* Chicago: University of Illinois Press.

Donovan, Josephine, and Carol J. Adams, eds. 2007. *The Feminist Care Tradition in Animal Ethics.* New York: Columbia University Press.

Goodall, Jane. 2000. Forward to *Rattling the Cage: Toward Legal Rights for Animals,* by Steven M. Wise, ix–xiii. Cambridge: Perseus.

Gruen, Lori. 2014. "Dignity, Captivity, and an Ethics of Sight." In *The Ethics of Captivity,* edited by Lori Gruen, 231–47. Oxford: University of Oxford Press.

———. 2015. *Entangled Empathy: An Alternative Ethic for Our Relationships with Animals.* New York: Lantern Books.

Haraway, Donna. 2003. *The Companion Species Manifesto: Dogs, People, and Significant Otherness.* Chicago, Prickly Paradigm.

Horta, Oscar. 2014. "The Scope of the Argument from Species Overlap." *Journal of Applied Philosophy* 31 (2): 142–53.

Kain, Patrick. 2009. "Kant's Defense of Human Moral Status." *Journal of the History of Philosophy* 47 (1): 59–101.

Kant, Immanuel. 1996. *Metaphysics of Morals.* Translated and edited by Mary McGregor. Cambridge: Cambridge University Press.

———. 1997. *Lectures on Ethics.* Edited by J. B. Schneewind. Translated by Peter Heath. Cambridge: Cambridge University Press.

Kim, Claire Jean. 2015. *Dangerous Crossings: Race, Species, and Nature in a Multicultural Age.* Cambridge: Cambridge University Press.

Kittay, Eva Feder. 2010. "The Personal Is Philosophical Is Political: A Philosopher and Mother of a Cognitively Disabled Person Sends Notes from the Battlefield." In *Cognitive Disability and Its Challenge to Moral Philosophy,* edited by Licia Carlson and Eva Feder Kittay, 393–413. Oxford: Wiley-Blackwell.

Korsgaard, Christine. 1996a. *Creating the Kingdom of Ends.* Cambridge: Cambridge University Press.

———. 1996b. *The Sources of Normativity.* Cambridge: Cambridge University Press.

———. 2004. "Fellow Creatures: Kantian Ethics and Our Duties to Animals." Tanner Lecture on Human Values, University of Michigan, February 6. https://dash.harvard.edu/bitstream/handle/1/3198692/korsgaard_FellowCreatures.pdf?sequence=2.pdf.

McMahan, Jeff. 2005. "Our Fellow Creatures." *Journal of Ethics* (9): 353–80.

———. 2008. "Challenges to Human Equality." *Journal of Ethics* (12): 81–104.

Midgley, Mary. 1983. *Animals and Why They Matter.* Athens: University of Georgia Press.

Mills, Charles. 1998. "Alternative Epistemologies." In *Blackness Visible: Essays on Philosophy and Race,* 21–39. Ithaca, NY: Cornell University Press.

Oderberg, David S. 2000. "The Illusion of Animal Rights." *Human Rights Review* 37: 37–45.

Pachirat, Timothy. 2011. *Every Twelve Seconds: Industrialized Slaughter and the Politics of Sight.* New Haven, CT: Yale University Press.

Rachels, James. 1990. *Created from Animals: The Moral Implications of Darwinism*. Oxford: Oxford University Press.

Regan, Tom. 1983. *The Case for Animal Rights*. Berkeley: University of California Press.

Regan, Tom, and Singer, Peter. 1989. *Animal Rights and Human Obligations*. Englewood Cliffs, NJ: Prentice Hall.

Ryder, Richard D. 1989. *Animal Revolution: Changing Attitudes toward Speciesism*. Oxford: Basil Blackwell.

Sanbonmatsu, John, ed. 2011. *Critical Theory and Animal Liberation*. Lanham: Roman and Littlefield.

Schlosser, Eric. 2002. *Fast Food Nation: What the All-American Meal Is Doing to the World*. London: Penguin Books.

Singer, Peter. 1975. *Animal Liberation*. New York: Harper Collins.

Wolfe, Cary. 2003. *Animal Rites: American Culture, the Discourse of Species, and Post-humanist Theory*. Chicago: University of Chicago Press.

Super-Muscly Pigs

Trading Ethics for Efficiency

Judith Benz-Schwarzburg and Arianna Ferrari

A nimal research is moving rapidly in two, divergent directions. Research on animal cognition, behavior, and welfare is teaching us that many animal species have complex cognitive and emotional lives and needs. Such insights reasonably increase our empathy for and insight into species other than our own. At the same time, new gene editing technologies are allowing scientists to design animals in ways that maximize their economic value as food sources. These technologies include programmable nuclease-based genome editing technologies, such as zinc-finger nucleases, transcription activator-like effector nucleases (TALENs), and the CRISPR/cas9 system. They permit the direct manipulation of virtually any gene of a living organism more easily, cheaply, and accurately than has ever been possible before. In the last five years, these technologies have been used to edit the germline of more than 300 pigs, cattle, sheep, and goats. In June 2015, a team of scientists from South Korea announced the creation of super-muscly pigs using the single-gene editing technology TALENs. Whereas the debates on the ethical and social aspects of genome editing of human embryos and crops have triggered public, political, and media firestorms, genome editing in animals has received virtually no ethical scrutiny. Yet gene modification of farm animals like super-muscly pigs raises complex ethical questions about animal welfare, about who is benefiting from these technologies, and about the evolving, contradictory relationship between humans and animals. These questions have been ignored so far, but our growing awareness of the rich inner lives of many animal species makes such neglect increasingly troublesome.

The Animals' Perspective

Concerns about the welfare of genetically engineered animals starts with the very process of creating them. Sperm and egg donors and surrogate mothers are normally killed if they are not "re-usable" for other purposes (such as in other animal experiments). Furthermore, animals whose modifications end up being undesirable either die because their health is severely compromised, or are killed because they are neither commercially valuable nor usable for scientific purposes. Due partly to the novelty of the methods, little data are available on the costs of new engineering techniques in terms of the total numbers of animals used, the unintended suffering created, and the effects that they are having on the phenotypes of various species. For the super-muscly pigs, the large size of the newborn piglet leads to birthing difficulties; only 13 of the 32 piglets created by the South Korean scientists survived as long as eight months, and only one survived considerably longer in a healthy state.

For sheep, goats, cattle, and pigs, genome editing techniques are used to modify primary cells, which are then transferred to the recipient mother via somatic cell nuclear transfer (cloning). This causes a range of animal welfare problems, such as very low live birth rates in some species; abnormal sizes, which render them incapable of natural movement; and respiratory and cardiac problems. Although genome editing techniques are expected to offer a more precise modification of the genome, they, too, generate many more animals than are actually used for experimentation. For example, a 2016 paper by Wenfang Tan and others in *Transgenic Research* surveyed the published literature and determined that out of 23,216 pig embryos, which were implanted in 112 pigs and generated 62 pregnancies, 237 pigs were born alive. Of these, 179 (76%) were properly modified or "edited," whereas the remaining 58 were not usable for the experiments. The scientists working in the field concentrate on the 76% of the pigs born alive and properly edited, which represents a success with respect to previous technologies. However, if we go back to the number of the embryos needed, to the pigs involved in pregnancies as well as to the individuals born which do not carry the modification needed, we can easily see how evaluation of the "efficiency" of these procedures depends on whether and how one counts the lives of the animals involved at all stages of the process. Among other considerations, as noted in a 2015 paper by Goetz Laible, Jingwei Wei, and Stefan Wagner, the "high efficiency" of CRISPR/Cas9 gene editing comes at the price of unintended mutations elsewhere in the genome.

For the scientists working on animal biotechnology in agriculture, the "efficiency" of gene editing provides a path to "solutions to securing food security for a rapidly growing human population under constrains of decreasing resources and a changing world climate," as Laible and colleagues explain. However, if we move the focus away from the perception of animals as sources of food products which need to be optimized, and instead consider the new types of costs that animals must pay for being created through these technologies, the picture becomes very different.

Threats to the animals' welfare do not end with their creation in the lab. Pigs, like many other animals used in agriculture, are being increasingly recognized by scientists as having complex abilities and needs in both the cognitive and social domains. As sentient beings, they have what biologists and veterinarians call "ethological needs," comprising, for example, the need to explore their surroundings and to engage in meaningful social interaction with others of their species. These needs are an innate and important part of the animals' behavioral repertoire. For example, animal welfare researchers have shown that if sows are not permitted to build nests prior to piglet birth, welfare problems such as abnormal repetitive behavior including bar-biting (chewing the metal bars of their crates), tail-biting, head-weaving or vacuum chewing (chewing when nothing is present) appear. Farm animal suffering manifests in diseases, lesions, or injuries (sometimes linked to high stock densities or quality of flooring), due to lack of space and behavioral stimuli, malnutrition, stress during handling, isolation, transportation, and, ultimately, killing methods.

We mention these well-known signs of animal discomfort because gene-editing of pigs and other farm animals is not being developed with consideration of how it might contribute to or even exacerbate such suffering. Especially in countries with very minimal or no animal welfare regulations, increased productivity due to genetic engineering could simply lead to more inhumane breeding, raising, and slaughter of ever-greater numbers of animals. Scientists who view genetically modified (GM) animals only in terms of food production fail to take into account the ethical costs of their one-dimensional perspective. The rationale for such a narrow view is obvious—increased yields in mass production of meat directly translate to increased profitability. But who is accounting for the increased animal suffering that gene editing technologies enable? Our concern is that the value gained from the higher precision of modifications compared to selective breeding, not to mention the economic interests involved in the meat market, create a powerful disincentive to consider the welfare concerns connected to GM animals.

The Global Context

The ambition behind the genetic modification of pigs and other animals bred for human consumption is to increase economic productivity. Scientists developing GM animals insist that their engineered status poses no threat to the environment, but such arguments cannot address larger questions of a globally sustainable food policy.

The livestock sector accounts for 14.5% of total global greenhouse gas emissions, which is more than the entire transport sector. It is the largest global source of the greenhouse gases methane (from ruminant digestive processes) and nitrous oxide (from manure and fertilizers used in the production of animal feedstuffs). The 2011 European Nitrogen Assessment estimated that in Europe, 85% of harvested nitrogen is used to feed livestock, with only 15% feeding people directly—even as the average European Union (EU) citizen consumes 70% more protein than needed for a healthy diet. Animal production creates substantial water and land pollution and requires vast amounts of territory—an estimated 45% of the global land surface area. Meat consumption is also linked to increased health risks such as cancer, ischemic heart disease, stroke, and diabetes mellitus. A report of the United Nations Environment Program (UNEP) concludes that both human and global environmental health would benefit from a substantial diet change away from animal products on a global scale. Yet demand for meat and dairy products continues to rise worldwide, driven especially by expected rising standards of living in China, India, and Russia. According to the Organisation for Economic Co-operation and Development (OECD)/Food and Agriculture Organization (FAO) Agricultural Outlook 2015, global meat production rose by almost 20% over the last decade and is expected to further expand until 2024. Pig meat production is expected to expand by 12% and poultry by 24% relative to 2012–14. The expansion is driven, in part, by increased profitability, particularly in these two sectors. Developing countries will account for approximately 75% of the additional output. To the extent that genetic modification of livestock increases meat production, it is also likely to lead to a "rebound effect," driving prices down and further increasing the demand for animal products—just as making roads wider or paving more parking lots tends to make traffic problems worse. Thus, economic incentives, changing demographics and dietary habits, and advances in gene editing technology are all pushing in the same direction toward increased stress on global environmental and food production systems.

Given these concerns, there is profound unresolved tension between, on the one hand, the plea by an increasing number of scientists and institutions like the OECD/FAO and the UNEP for a substantial reduction in agriculture's adverse impact through a decrease in the use of animals for food, and, on the other hand, biotechnologists' support for ever-enhanced forms of animals for meat and dairy production. This tension is fundamentally political in nature, yet it is also a problem of ethics and values, not only because these technologies have negative impacts on animal welfare, but also because they profoundly shape the way in which we think about animals.

The Changing Human-Animal Relationship

Increased knowledge about animal sentience, cognition, and behavior is changing the human-animal relationship. Decades of ethical debates and a tightening of animal welfare regulations in many countries around the world contributed to higher awareness of the needs of some animals—predominantly the ones we take into our homes. We increasingly attribute emotions, intelligence, needs, and even some rights to our companion animals. We usually do not kill and eat them. Our perception of farm animals, like pigs, is substantially different, of course. We only pet the ones we love and tend to forget about the ones we eat—a cognitive dissonance which to us is one of the most striking and ethically problematic features of today's human-animal relationship. Psychological studies show, unsurprisingly, that we tend to resolve our cognitive dissonance by denying the suffering and complex cognition of the species we want to eat. Meanwhile, scientific evidence increasingly shows that farm animals share capacities like sentience and cognition with humans, thus making it increasingly difficult to grant them an exemption from moral relevance.

Although science aimed at understanding the cognitive capacities of animals gives us reason to appreciate and empathize with dogs and pigs alike, technological advances like gene editing enhance the drift of the moral status of farm animals away from living beings that have inherent value (a moral standing in themselves) to products which only have instrumental value and a market price. Gene-edited animals like super-muscly pigs are brought to life and designed to meet human desires. Scientists speak of eliminating undesired traits to enhance productivity. In the process, it may even be the case that these designed animals suffer more than their non-GM relatives because of their extreme physiological traits. They are a technological success exactly *because* they are optimized—for example via an increase in body mass to match their instrumental purpose—but what about their suffering?

The scientific creation of super-muscly pigs thus drives a problematic understanding of animals. It does so in times where we are increasingly aware that animals, at least sentient animals, have needs and interests themselves and objective attributes of individual welfare, and thus moral standing that is independent from their instrumental value as objects for our consumption.

As such, the application of gene editing to animals, which to some is a step forward for technology, may be a step backwards for human moral development, as it conceals and heightens conflicts and dilemmas about which a truly reflective society should be openly deliberating. The continued pursuit of our scientific capacity to engineer animals for consumption, without commensurate attention to the ethical issues at the center of global meat production is at best naive, and at worst irresponsible. Recent decades have seen significant strides in confronting the ethical challenges of our relationships with animals, progress that is supported by our increasing scientific knowledge about animal cognition and sentience. Concern for animal welfare is rising. Meanwhile, from the overuse of antibiotics to the pollution of waterways from feedlots, the environmental consequences of the global food system add another dimension of concern to the globalizing market for meat. In all, our growing understanding of the human-animal relationship suggests that we need to revise our view of sentient animals, such as pigs, to recognize that far from being merely editable genetic material and edible flesh, they are also living individuals that merit our serious moral consideration.

DISCUSSION QUESTIONS

1. How does the notion of "person" affect protections in research?

2. How could animals be protected with similar research ethics?

3. What rules/principles would you recommend for a code of research ethics?

4. What justice could be done for victims of unethical research (e.g., Henrietta Lacks, individuals from Tuskegee, etc.)?

5. What should be done about the propagation of research that is known to be false?

6. How do codes of research ethics differ by the ethical theory we apply? In what ways are they similar?

7. How could a global ethical standard of research be enforced?

8. If research is completed unethically, should the research still be published if the information gathered has the potential for positive impact?

Recommended Cases

Edward Taub—In 1981, Edward Taub studied somatosensory deafferentation in monkeys by surgically cutting all the nerves in one limb and trying to stimulate regrowth. Taub believed that the voluntary nonuse from stroke damage may be adapted using the psychological principle of learned helplessness. PETA took photographs of the monkeys in Taub's lab, which showed them in pain and poor conditions. Taub was convicted of failing to provide proper veterinary care for not bandaging wounds. The monkeys were removed from Taub's lab. In 1990, an experiment was done on the monkeys from Taub's experiment. It was discovered that their brains had been rewired—proof of neural rewiring, the "Holy Grail of rehabilitative medicine."

MMR Vaccine and Autism—Despite an uncontrolled design, a small sample size, and speculative conclusions, Wakefield's study positing a link between autism and the MMR vaccine received vast public interest. More thorough studies were then completed and showed that there is no link between the MMR vaccine and autism. Ten of the 12 authors of the study retracted their involvement with the paper, and it was discovered that Wakefield was being paid to conduct the research by lawyers representing parents in a lawsuit against vaccine-producing companies.

Stanford Prison Experiment—In 1971, Robert Zombardo conducted an experiment in which student volunteers were randomly assigned roles as prison guards or prisoners. After a mock arrest, the prisoners were brought to the basement of the psychology building, stripped of their clothes, and assigned a number for a name. Due to the mental torment which the prisoners underwent at the hands of the prison guards, multiple mental breakdowns occurred, and the experiment was cut short early. It is one of the most influential experiments in understanding roles, symbolism, situational validation and group identity. The findings were especially prescient in regard to the atrocities at Abu Ghraib.

UNIT III

Eugenics

....

Unit III Introduction

Coined by Sir Francis Galton, the term "eugenics" comes from two Greek roots: "eu," meaning good or well, and "genea," meaning decedents, birth, generation. Eugenics tries to achieve "good descendants" or "good genes," or, as Galton called it, "the science of improving human heredity over time." Galton sought to achieve this by having political and social structures mirroring this goal. In 1933, eugenics was called "an exceptionally important public health initiative ... we go beyond neighborly love; we extend it to future generations" by the counselor of the Reich Interior Ministry.

While eugenics seems to be an issue of the past and of faraway lands, this is untrue. The case of *Buck v. Bell* (1927), a Supreme court ruling that has never been overturned, confirmed the legality of forced sterilizations of inmates/patients with hereditary imbecility or insanity. Oliver Wendell Holmes, writing for the majority decision, wrote that "It is better for all the world, if instead of waiting to execute degenerate offspring for crime, or to let them starve for their imbecility, society can prevent those who are manifestly unfit from continuing their kind ... Three generations of imbeciles is enough." In the state of Connecticut for instance, an individual could have been forcibly sterilized if they met the criteria of "feebleminded, insane, idiots, imbeciles, and those with inherited tendency to crime." In our upcoming unit on medicalization (Unit VIII), we shall see how loosely these terms could be used for eugenical purposes. Connecticut's sterilization law was enacted in 1909 and changed to sterilization becoming voluntary in 1965, one of America's shortest state-sponsored eugenics programs.

Eugenics is most often carried out in two ways. The first method is to encourage or require the ideal individuals to have children with other people of their ilk. The second method is to make undesirable individuals unable to reproduce. Peoples have used these methods to ensure that desirable traits continue to be passed down while attempting to weed out undesirable traits. Notions of "desirable people" carry through as high abortion rates based on gender, indicators of Down syndrome, and other factors point toward a difficult balance between autonomy and maleficence. It is this balance we will explore in this unit.

In "Eugenics," we will explore the history of eugenics as pseudoscience and public policy. The selection emphasizes the American eugenics policies, the ways they informed Germany's Nazi regime, and the ways American eugenics was put into practice. In addition, we will see the impact of these policies as it relates to racism, xenophobia, and nationalism.

"Is Selective Termination Morally Wrong?" specifically grapples with termination decisions surrounding indicators of Down syndrome. Beginning with a section titled "Bias as Cause," we are confronted with assumptions about the lives of individuals with Down syndrome and compare those with data about how individuals with Down syndrome feel about their lives. Also discussed is the harm that is done to communities of individuals when termination is chosen based on the traits of those communities.

The role of medical professionals is the focus of "Informed Consent or Institutionalized Eugenics? How the Medical Profession Encourages Abortion of Fetuses with Down Syndrome." The authority of doctors, geneticists, and other caretakers is valued by individuals in vulnerable positions. Biased presentation and perceived legal liability can make giving truly informed consent difficult. By examining how practitioners communicate, as well as the veracity of information given, women can make truly autonomous decisions.

From Darwin to Galton

Hallam Stevens

A lthough eugenics was developed and implemented in many countries, the roots of eugenic ideas are usually ascribed to the work of the nineteenth-century naturalist Charles Darwin (1809–1882). In his *On the Origin of Species* (1859), Darwin proposed the theory of natural selection: organisms evolve through competition with one another—those best adapted to their environment survive while some species become extinct (sometimes called "survival of the fittest"). Darwin's ideas had certainly been influenced by his observations of Victorian society. He had seen or read about England's poor dying in workhouses and slums, and he had watched some of his own children perish due to illness. But, at first at least, Darwin did not dare extend his theory to take account of human societies—its suggestion that humans were distantly related to apes was controversial enough as it was.

But others were not so reticent. Darwin's work was a revelation to many, explaining so much of the complexity of nature with one simple principle. For some, its implications for society were immediate and obvious. The individual who took up these ideas most enthusiastically was Francis Galton (1822–1911), Darwin's own cousin. Beginning in 1865, Galton began to publish his ideas about applying Darwinian theory to humans. Galton believed that by selecting "better" or "fitter" human beings for marriage (and reproduction), society could gradually be improved. After all, farmers and breeders selected the best cows or the best dogs for breeding in order to produce the most desirable offspring. Couldn't the same principle be applied to humans? Galton believed it could: it is "quite practicable," he wrote, "to produce a highly gifted race of men by judicious marriages during several consecutive generations."[1]

For this scheme to make sense, however, Galton had to show that socially valuable human traits (such as intelligence) were heritable (passed on from parents to offspring). If traits weren't heritable, then selection would have no effect.[2] Intelligence, however, was difficult to measure. Sensibly, Galton

---

1 Francis Galton, *Hereditary Genius: An Inquiry into Its Laws and Consequences* (London: Macmillan, 1869), 1.

2 If intelligence was heritable, then we would expect intelligent parents to produce intelligent children. If intelligence was not heritable, then we would expect intelligence to be randomly distributed through the population—in other words, anyone could be intelligent, regardless of parentage.

began by attempting to show that other, simpler, human traits were passed down from generation to generation. Galton set up a laboratory in London to measure the heights, weights, arm spans, and breathing capacity of thousands of individuals. The Anthropometric Laboratory, established in 1885, was especially interested in comparing measurements across generations: between grandfather, father, and son, for instance. If Galton could show that taller grandfathers produced taller fathers and taller sons, then it would strongly suggest the heritability of height, for instance.

Galton's work in this field had a lasting legacy: the development of modern statistics. Galton realized that not all tall fathers have tall sons—there was not an *absolute* or *deterministic* relationship between generations. Rather, it was perhaps only that tall fathers had a greater *likelihood* of producing sons taller than *average*. To express these relationships, Galton (and his followers) developed statistical techniques such as *correlation* and *regression*. Galton showed that traits such as height were distributed according to a *normal curve* and that selecting shorter or taller parents shifted the shape of the curve for the subsequent generation.

In the end, Galton also tried to show that the same sorts of relationships applied to less quantifiable traits. In the 1860s, Galton began compiling the family histories of "eminent men" in England (such as Fellows of the Royal Society). Galton's hypothesis was that eminence was passed through families: eminent ancestors produced eminent descendants. Galton published the results of his study in what he called "historiometry" in his 1869 book *Hereditary Genius.*

Galton's work was continued by his protégé, Karl Pearson (1857–1936). Pearson was a skilled mathematician and further developed Galton's statistical ideas, gaining increasing scientific credibility for the new field of eugenics. Pearson inaugurated the journal *Biometrika* in 1902 to publish eugenic ideas. In 1911, he took up the specially created post of Galton Eugenics Professor at the new Department of Applied Statistics at University College London. By the first decades of the twentieth century, eugenics comprised a well-respected and widely supported set of scientific ideas.

The plausibility and prestige of eugenics was enhanced by the ways in which it resonated with European (and especially English) notions of race and national competition. The British Empire, in particular, was justified on the grounds of racial superiority: the English ruled over Indians and other peoples because of their superior racial characteristics, Victorians believed. "History shows me only one way," Pearson espoused, "in which a high state of civilization has been produced, namely the struggle of race with race, and the survival of the physically and mentally fitter race."[3] But if international politics was determined by racial characteristics, then this meant that preserving those characteristics was absolutely essential for maintaining power.

If inferior individuals, especially the poor and the lower classes, reproduced more frequently than the better types, the English race would quickly degenerate, eugenicists believed. The result would be catastrophic: Britain would lose its special status in the world order. This "degeneration" is exactly what many eugenicists saw around them. Galton and Pearson noted with dismay that the poor *did*

---

3 Karl Pearson, *National Life from the Standpoint of Science* (London: Adam & Charles Black, 1901), 19–20.

seem to be having more offspring than the upper classes. Many eugenic policies were aimed at averting this national disaster by discouraging the reproduction of the poor and encouraging the reproduction of the rich (Pearson suggested giving economic incentives to well-matched couples to have babies).

Galton and Pearson's work attracted much attention in Britain, especially amongst the elite. But eugenic policies were never implemented by the British state. Other countries—including Argentina, France, Germany, the Soviet Union, and the United States, and Scandinavian nations—not only produced their own eugenic ideas but also implemented them to various degrees before World War II. Many of these nations experienced rapid population growth (especially through migration) in the last decades of the nineteenth century. This expansion was accompanied by an increasing role for the state in providing health care. These two conditions increased the perceived legitimacy of state intervention to safeguard the health of the population.

In France, for instance, a French Eugenics Society was founded in 1912 and a National Social Hygiene Office was founded in 1924 (funded by the Rockefeller Foundation). The latter focused on measures such as treating tuberculosis, preventing alcoholism and prostitution, and controlling sexually transmitted infections. In other words, eugenics was limited to attempts to combat disease and improve living conditions. In Sweden, on the other hand, eugenic ideas led to the implementation, in 1934, of a compulsory sterilization law. The law sterilized persons deemed "unfit" to have children due to perceived mental, physical, or social deficiencies. In the United States and Germany too, eugenics moved from "positive" (attempts to encourage procreation amongst "healthy" individuals) to "negative" (attempts to stop or outlaw procreation of "unhealthy" individuals).

Eugenics in the United States

It was in the United States that eugenics found its way most deeply into social policy. America's history of slavery and immigration made questions and problems of race (and especially race mixing) especially pertinent. The second half of the nineteenth century saw not only the emancipation of African Americans, but also increased levels of immigration of Irish, Chinese, Italians, and Jews. As in Britain, this population explosion caused increasing anxiety about the potential dilution of racial strength.

The most influential American convert to eugenics was Charles Davenport (1866–1944). Davenport was inspired by the rediscovery of Mendel's laws in 1900. In the 1860s, an obscure Austrian monk— Gregor Mendel—had conducted a series of experiments on peas in his vegetable garden. His results showed that specific traits of the peas (such as color) were passed on from generation to generation according to precise numerical ratios. This theory, widely circulated only long after Mendel's death, seemed to support eugenic notions that specific traits were passed on from generation to generation.

In 1904, Davenport persuaded the Carnegie Institute of Washington to provide funds to establish an experimental station at Cold Spring Harbor on Long Island, New York. The Station for Experimental

Evolution was dedicated to searching for Mendelian traits in humans. Davenport used the generous funding to begin collecting extended family pedigrees from around the United States. These recorded the incidence of mental diseases, "imbecility," epilepsy, criminality, "sexual immorality," alcoholism, and other undesirable traits. Like Galton, Davenport aimed to show that these characteristics were passed down through families. This was painstaking work, resulting in a widely circulated and acclaimed book, *Heredity in Relation to Eugenics*, published in 1911.

Such work was expected to have significant social implications. In particular, it showed, according to Davenport, that the mixing of races (especially between white and black) was producing inferior offspring and gradually weakening American society (figure 3.1). "The great influx of blood from Southeastern Europe," Davenport worried, would make the American population, "darker in pigmentation, smaller in stature, more mercurial, ... more given to crimes of larceny, kidnapping, assault, murder, rape, and sex-immorality."[4] In the first place, Davenport recommended stricter rules for immigrants, checking their family history to ensure that the "nation's protoplasm" was not compromised. These eugenic ideas found expression in the Immigration Act of 1924. The new law created quotas based on national origin. It severely limited immigration from Southern and Eastern Europe (including Jews), and prohibited immigration from the Middle East, East Asia, and India.

Figure 3.1 Color inheritance in guinea pigs (ca. 1925). Exhibit designed to educate the public about Mendelian inheritance and eugenics. Reproduction between two white guinea pigs produces healthy offspring. Reproduction between two black guinea pigs produces black guinea pigs. Reproduction between black and white guinea pigs, however, produces smaller and seemingly deformed offspring. Such displays were supposed to demonstrate the eugenic danger of racial mixing. Source: American Philosophical Society.

---

4 Charles Davenport, *Heredity in Relation to Eugenics* (New York: Henry Holt, 1911), 219.

As eugenics grew in scientific reputation and popularity between 1910 and 1920, other policies began to be implemented too. "Race hygiene," as it was known in some contexts, began to become a worldwide movement with prominent scientist and politicians offering their support. In the United States, eugenic laws were largely adopted on a state-by-state basis. Initially, these efforts focused on encouraging appropriate marriages between "fit" individuals. For instance, some states provided prizes for "fitter families" or "better babies"—those certified as fulfilling eugenic criteria (figure 3.2). Competitions were held at state fairs and ads were run in newspapers. Education was a key component of these campaigns: eugenicists attempted to inform the public about how to spot a eugenically fit partner and warn people about the potential consequences of "bad matches."

Figure 3.2 Better Babies Contest award certificate. Louisiana State Fair, Shreveport, 1913. Certificate from a contest designed to encourage matches and reproduction between eugenically fit individuals. Source: Mrs. Frank deGarno Papers, MS#1879, University of Tennessee Libraries, Knoxville, Special Collections.

Some of this now seems rather far-fetched. We must not forget, however, that biological fitness was understood as the cause of crime and disease and therefore a critical social problem. Moreover, prominent scientists and scientific institutions of the time backed it up. However racist and misguided they now appear, many eugenicists sincerely believed that they were acting in society's best interest, protecting it from degeneration and ruin.

By the 1920s, eugenic measures in some jurisdictions began to go even further. Instead of just encouraging beneficial matches they began to actively intervene to prevent "bad" matches. The first step was marriage laws. As many as thirty states adopted laws that restricted marriage if either party was considered eugenically "unfit" (often using such criteria as "feeble-mindedness" or evidence of sexually transmitted disease).

For some eugenicists, even this did not go far enough. For one thing, unmarried people could have babies anyway, and thereby still affect the overall gene pool. One solution was to remove this possibility completely. The first eugenic sterilization law was passed in Indiana in 1907. But by the end of the 1920s, thirty-one states had enacted laws that forced some types of "unfit" men and women to undergo compulsory (involuntary) sterilization. In a 1927 case known as *Buck v. Bell*, the US Supreme Court declared such laws constitutional. Although men were sterilized to reduce aggression and control criminal behavior, many of the laws were directed at women as child-bearers, holding them more accountable for reproduction (about 60% of all sterilizations were performed on women).

Those subject to sterilization varied from state to state. Mostly commonly the legislation specified "imbeciles" or others deemed mentally deficient. In some cases, this extended to the mentally ill, the deaf, the blind, epileptics, and individuals with physical deformities. The racialized aspects of eugenic thinking became all too clear: African Americans and Native Americans became particular targets. Many individuals were sterilized not only against their will, but also sometimes without their knowledge (one practice was to perform sterilizations when a patient presented at a hospital for another condition or operation, including childbirth).

Between 1907 and 1963, about 64,000 forcible sterilizations were performed in the United States. Many of the sterilization laws remained in place until the 1960s and '70s, although the number of sterilizations declined from the 1940s onward.

Eugenics in Germany

We usually associate the policies of Nazi Germany with unmitigated racism. Indeed, Adolf Hitler's racism targeted the Jews for complete genetic extermination. However, the Nazi regime also sought to expel or kill homosexuals, the mentally ill, the physically handicapped, and the otherwise sick and infirm. These Nazi racial and social policies were informed by eugenic ideas.

However, eugenics was not imported into Germany by the Nazis—its theories had long and deep roots amongst German thinkers. Eugenics in Germany was developed through the work of Alfred Ploetz (1860–1940) and Wilhelm Schallmayer (1857–1919). In his 1895 book *The Efficiency of Our Race and the Protection of the Weak*, Ploetz described an ideal society run according to eugenic principles: marriage and reproduction would be limited according to moral and intellectual capacities and disabled, sick, and weak children would be killed. Schallmayer likewise believed that eugenic policies were the key to solving social problems (such as overcrowding, epidemics, poverty, and crime) and achieving national power for the newly unified German state.

German defeat in the First World War fueled fears of the decline of the "Nordic race" and boosted eugenic thinking. In the 1920s, however, Germans looked more and more to the United States for examples of how to deal with racial and eugenic problems. For the most part, the US was happy to help. In 1927, the Rockefeller Foundation provided funds to establish the Kaiser Wilhelm Institute for Anthropology, Human Heredity, and Eugenics in Berlin. The work of its director, Eugen Fischer (1874–1967), influenced Hitler's racial thinking.

Once the Nazis came to power in 1933, they immediately began to implement their eugenic policies. In July 1933, the Reichstag passed the "Law for the Prevention of Hereditarily Diseased Offspring," requiring physicians to register known cases of hereditary disease. By 1934, the Nazis had implemented laws permitting involuntary sterilization for feeble-mindedness, mental illness, epilepsy, and alcoholism. Nazi propaganda emphasized the great cost to the state and to society of caring for physically and mentally ill people (figure 3.3). German eugenic laws were closely modeled on US examples and the Nazis closely followed eugenic developments in America. The Nazis propaganda also emphasized that Switzerland, Britain, Japan, Scandinavian countries, and the United States all had similar eugenic laws: "We do not stand alone," Nazi eugenic posters proclaimed.

Figure 3.3 Nazis and eugenics. Comparison of daily living costs for an individual with a hereditary disease and for a healthy family (from the series "Blood and Soil," ca. 1935). The left-hand texts reads: "An invalid costs the state RM5.50 per day"; the right-hand reads: "For RM5.50 a hereditarily healthy family can live for one day!" This poster was part of the National Socialist effort to improve the German people through "racial hygiene." Source: Bildagenter Preussische Kulturbesitz.

In 1935, the infamous Nuremberg Laws prohibited marriages between Aryans and "unfit" persons. Jews, of course, were the main targets, but other eugenically unfit persons also fell under the scope of the law. By the outbreak of war in 1939, some 400,000 individuals had been sterilized. As the war created a growing demand for hospital beds and medical resources, sterilization was not enough: a "final solution" was required for those who were deemed to be a burden on German society. For the Jews, this meant a policy of total extermination. Other groups too—Roma, homosexuals, communists, and Slavs—were targeted for eugenic, racial, and political reasons. Between 1939 and 1941, doctors murdered 70,000 mental patients at psychiatric hospitals, gassing them with carbon monoxide. In 1941, many of these doctors were reassigned to concentration camps to assist with the mass murder of other "undesirables."

Bias as a Cause

Chris Kaposy

....

As we have seen, termination rates after a prenatal diagnosis of Down syndrome are in the range of 60 to 90 percent (Natoli et al. 2012), which means that large majorities of prospective parents do not desire an outcome in which their child has Down syndrome.

In a number of empirical studies (Bryant, Green, and Hewison 2010; Choi, Van Riper, and Thoyre 2012), researchers have asked pregnant and nonpregnant women about what they would do if given a prenatal diagnosis of Down syndrome, and the reasoning behind these hypothetical choices. One study has investigated the reasons that women give for their actual decision to terminate after a prenatal diagnosis of Down syndrome (Korenromp et al. 2007). In this study, originating out of the Netherlands, the women who had terminated their pregnancies after a prenatal diagnosis of Down syndrome identified a series of motivations, which can be organized into two groups. One group of motivations comprised the beliefs, fears, and desires that the respondents had about themselves or their other family members. More than 60 percent of the women stated that

- they considered the burden of having a child with Down syndrome too great for their other children;
- they considered the burden too heavy for themselves;
- they did not want a disabled child;
- they thought they would become unhappy with having this child (Korenromp et al. 2007).

The other group of motivations comprised beliefs, fears, and desires that the respondents had about their affected fetus, understood as a prospective child. More than 80 percent of the women stated that

- they believed that the child would never be able to function independently;
- they considered the abnormality too severe;
- they considered the burden for the child himself/herself too heavy;
- they worried about the care of the child after their own death (Korenromp et al. 2007).

In one of the studies involving hypothetical decision making, researchers found that beliefs about negative quality of life for parents who have a child with Down syndrome and negative attitudes toward people with Down syndrome, were significantly associated with their intention to terminate an affected pregnancy (Bryant, Green, and Hewison 2010). In this study, the term "negative attitude" had a broad definition and included any emotion such as sadness or pity that could be assigned a negative valence.

The great unmentionable and largely undetected factor in these and other empirical studies is the influence of attitudes of bias against people with Down syndrome in the decision making of prospective parents. No one would be willing to admit that they hold a bias against people with Down syndrome. People are usually able to find an interpretation of their attitudes that does not appear to have the connotation of bias. But bias can have an unconscious influence and its effects can be detected in the motivations that people are willing to acknowledge. Attitudes of bias can be implicit and those holding these attitudes may be unaware of them.

Bias is perhaps closest to the surface in the claim that "I did not want a disabled child"—offered by 63 percent of respondents in the study of women who had terminated their pregnancies after learning the fetus had Down syndrome (Korenromp et al. 2007). This statement expresses a rejection of the disabled prospective child simply because of the child's disability. The disability is a stigmatized trait. In one sense of the word, a stigma is an undesirable characteristic that marks someone as apt for rejection, or as someone who should be denied association. In this case, the stigmatizing trait of disability can change a wanted pregnancy into an unwanted one as soon as it is detected through prenatal testing. There are similar suggestions of bias in the claim that "I thought I would be unhappy with having this child"—offered by 61 percent of respondents in that same study (Korenromp et al. 2007). When this reason is offered, awareness of the single stigmatized trait of disability makes prospective parents unhappy with a previously wanted pregnancy.

Another form of bias can be detected in some of the motivations that relate to the prospective child's life. Bias can be detected in beliefs that a child with Down syndrome will not be able to live independently, that his or her disability will be too severe, or that the burden of life itself will be too great for the child. Instead of seeing the future child as a whole human being who will enjoy all kinds of possibilities for happiness and flourishing, prospective parents adopt a narrow view in which the potential child is viewed exclusively through the lens of a diagnosed disability. As Asch and Wasserman (2005, 181) argue, this kind of motivation for selective termination exemplifies an "uncritical reliance on a stigma-driven inference from a single feature to a whole future life." This is the "synecdoche" argument that testing and termination result from stereotypical generalizations about the potential child based on one piece of information provided by a test. Synecdoche is the literary device in which a part of something represents the whole. It is a form of stereotypical thinking to assume that "this one piece of information suffices to predict whether the experience of raising that child will meet parental expectations" (236). The biased "synecdochical" feature of this reasoning lies in the fixation on the disabled characteristic of the prospective child. The bias leads prospective parents to believe that a child with Down syndrome will necessarily lead a bad life or will be a disappointment because of Down syndrome.

Stereotypes tend to distort perceptions. Some of the common motivations for selective termination reflect inaccurate assumptions about living with Down syndrome or parenting a child with Down syndrome. In the empirical study I have been discussing, 83 percent of respondents who had terminated were motivated by a belief that Down syndrome would be excessively burdensome for the prospective child (Korenromp et al. 2007). In contrast, the study by Brian Skotko, Susan Levine, and Richard Goldstein (2011c; discussed in earlier chapters) reports that 99 percent of people they surveyed who are living with Down syndrome are happy with their lives.

Among respondents who had terminated in the Netherlands study, 73 percent believed that the burden of having a child with Down syndrome would be too great for their other children (Korenromp et al. 2007). Again, in contrast, research involving parents of children with Down syndrome shows that 95 percent of parents with other children say that their children with Down syndrome have good relationships with their nondisabled siblings. More than 90 percent of the nondisabled children themselves say they have feelings of affection for and pride in their siblings with Down syndrome (Skotko, Levine, and Goldstein 2011a).

In the interviews with women who had undergone selective termination, 64 percent had terminated because they felt the burden of bearing a child with Down syndrome was too great for themselves (Korenromp et al. 2007). But research with parents of such children tells a different story. More than 97 percent reported feelings of love and pride in their children with Down syndrome, and only 4 percent expressed regret about having such a child (Skotko, Levine, and Goldstein 2011b). The divergent findings of these studies suggest that perceptions about parenting a child with Down syndrome are distorted by stereotyped ways of thinking and thus also by bias. According to their reading of the evidence, Asch and Wasserman (2005, 175) argue that: "The most that can plausibly be claimed is that being or having a child with a disability is at times different and more difficult than being or having a 'normal' child, and that specific impairments are very unlikely to meet specific parental expectations (e.g., a child with Down syndrome is not likely to become a great mathematician like her mother)."

Table 3.1 summarizes the divergent findings.

Table 3.1 Comparison of perspectives on Down syndrome

	Perspectives of women who have selectively terminated because of a diagnosis of Down syndrome (Korenromp et al. 2007)	Perspectives of people living with Down syndrome, parents of children with Down syndrome (Skotko, Levine, and Goldstein 2011a, 2011b, 2011c)
Quality of life of child with Down syndrome	83%—Down syndrome would be excessively burdensome for the child himself/herself	99% of people with Down syndrome are happy with their lives
Effects on siblings of having a brother or sister with Down syndrome	73%—a child with Down syndrome would be too great a burden for his/her siblings	95% of parents report that children without Down syndrome have good relationships with their siblings who have Down syndrome
Effects on parents of having a child with Down syndrome	64%—having a child with Down syndrome would be too great a burden for myself	97–99% of parents express love and pride toward their children with Down syndrome

Perhaps this line of argument is unfair and the reasoning of prospective parents should be accepted at face value—which is to say that there is no bias against people with Down syndrome in these decisions. One could interpret the empirical studies of people who opt to terminate a pregnancy affected by Down syndrome to mean that women and their partners simply make decisions in accordance with what they perceive to be best for their families and for themselves. But there is another compelling argument that bias against Down syndrome must have an effect on decision making about termination for Down syndrome.

First of all, there is good evidence that widespread bias against Down syndrome exists. There are some signs that things are getting better, such as the Down syndrome self-advocacy movement. Nonetheless, one still regularly hears people use the words "retard" or "retarded" to signify someone or something that is stupid, broken, incompetent, or disliked. Recent Hollywood movies such as "The Hangover" (2009), "The Change-Up" (2011), and "Ted" (2012) have made people with cognitive disabilities the butt of jokes through the gratuitous use of the word "retard" or plot devices making fun of people with Down syndrome. These days, no Hollywood script would see the light of day if it had similar derogatory material about racial or ethnic groups, yet it still seems acceptable to make fun of people with cognitive disabilities. The currency of the "retard" slur in our culture has given rise to a countermovement to end to use of the "r-word," including a website listing thousands of people who have made public pledges to "eliminate the derogatory use of the r-word" (http: //www.r-word.org). Such a movement would not be needed if bias and stigma did not exist.

Furthermore, social science research into the implicit attitudes that people hold toward those with cognitive disabilities is strongly suggestive that negative attitudes are the norm (Enea-Drapeau, Carlier, and Huguet 2012; Hein, Grumm, and Fingerle 2011; Proctor 2012; Robey, Beckley, and Kirschner 2006; Wilson and Scior 2015). In a review article summarizing several studies in this area, Michelle Clare Wilson and Katrina Scior (2014, 315) conclude that all of the studies in their review show "moderate to strong negative implicit attitudes towards individuals with [intellectual disabilities]." The reviewed studies include one that specifically investigates implicit attitudes toward people with Down syndrome (Enea-Drapeau, Carlier, and Huguet 2012). Implicit attitude testing uses methods to determine how strongly people associate pairs of concepts, such as "disabled" and "terrible," "nondisabled" and "pleasant," or vice versa. Such tests can also measure the association between positive and negative concepts and visual images, such as the faces of people who have Down syndrome (Enea-Drapeau, Carlier, and Huguet 2012). The degree to which response times diverge between positive and negative attributions gives an indication of subconscious association between concepts and images (Wilson and Scior 2015).

These implicit attitude findings tend to contrast with findings that measure the *explicit* attitudes that people hold toward those with cognitive disabilities. Explicit attitude studies tend to find that people have very positive explicit attitudes toward this group (Ouellette-Kuntz et al. 2010; Wilson and Scior 2015). However, researchers suspect that explicit attitude research methodologies, such as surveys, can be complicated by the fact that research participants might be influenced by the desire to be viewed favorably by others, and thus may give answers that are more socially acceptable, rather than answering honestly (Ouellette-Kuntz et al. 2010). This problem has prompted the move to other methodologies, such as those used in implicit attitude research. There is also a study using a qualitative methodology showing that explicit positive attitudes toward people with cognitive disabilities can coexist with, or mask, attitudes that are more actively hostile (Coles and Scior 2012). These findings about implicit negative attitudes toward people with cognitive disabilities are consistent with the idea that bias against people with Down syndrome is common as well as unacknowledged or unrecognized.

Second, it seems highly unlikely that widespread bias against stigmatized disabilities like Down syndrome exists but does not affect decisions about carrying a fetus, giving birth to and parenting a child with the stigmatized trait. The high termination rates for fetuses diagnosed with Down syndrome do not indicate that prospective parents are immune to common cultural attitudes toward Down syndrome. Rather, we would expect a high termination rate for fetuses exhibiting a trait that is widely stigmatized in our culture. Of course, there are sure to be exceptions to the norm: individual cases in which women selectively terminate for Down syndrome for reasons innocent of bias. But, looked at from the perspective of the whole population of prospective parents who decline to parent a child with Down syndrome, the pervasive bias against cognitive disabilities must affect many of their decisions.

To be clear, I am not arguing that all decisions to selectively terminate after a prenatal diagnosis of Down syndrome are necessarily motivated by bias. Furthermore, it would be difficult to quantify motivation by bias such that one could argue that most selective termination decisions have this motivation. I have not done this. Instead, I have argued, first, that bias can be detected in the most common reasons offered by women for their selective abortion of a fetus with Down syndrome. Second, I have argued that since bias against people with Down syndrome is common in our culture, it must also be common in selective termination decisions. Neither of these arguments could successfully characterize all individual decisions to selectively terminate as biased decisions. These arguments are effective, however, in showing that bias against people with Down syndrome is an influence—even a significant influence—from a population-wide perspective on the high rates of selective termination. These high rates are the target of the disability critique, as I have formulated it.

The Nature of the Harm

At this point in the argument, it is unclear how selective termination is linked with any recognizable negative effects on anyone. The disability critique is advanced as part of an intramural debate between groups that are pro-choice. Because of these pro-choice commitments, proponents of the disability critique would not allege that selecting against Down syndrome results in wronging or harming any fetuses. According to the pro-choice view shared by those advancing the disability critique, the fetus is not a subject who can be wronged in this way. If the fetus cannot be harmed, that leaves the population of people living with Down syndrome as the group that must be suffering from the negative consequences of the high rates of selective termination (if there is any moral wrong involved).

But defenders of prenatal testing and selective termination have been skeptical of any putative link between these practices and harm or wrong done to people with Down syndrome. Jamie Lindemann Nelson (2007a, 476) points out that, "It will be difficult to provide good evidence for the counterfactual 'Had there been no social practice of prenatal testing and termination, people with disabilities would, other things being equal, have experienced less stigmatization.'" Nelson is right that selective termination in itself does not cause stigma. As I have been arguing, stigma and biased ways of thinking have actually been a primary cause of the push for prenatal testing and the high rates of termination, rather than being an *effect* of these practices. If so, then the common social practice of testing and termination for Down syndrome, and any hardships experienced by people with Down syndrome are not related as cause and effect, but rather both are effects of a common cause—stigma and biased attitudes toward Down syndrome. In a subsequent section, I will explore the basis for objecting to the high rates of selective termination for Down syndrome, given that this practice is not a cause of bias against people with Down syndrome. This section discusses one aspect of the harm experienced by people with Down syndrome resulting from the bias directed at them.

People with cognitive disabilities live within an atmosphere of bias. There are similarities between the effects of this bias on people living with Down syndrome and the way that this bias is manifest in prenatal decision making. The parent-child relationship is among the most intimate relationships in which we may ever be involved. The desire to avoid being a parent of a child with Down syndrome can thus be interpreted as a refusal to be involved in an intimate relationship (i.e., parenting) with a person who has Down syndrome.[1] Similarly, recent Canadian research into the lives of young people with various disabilities reveals a parallel form of this bias. One study has found that more than half of the young people with cognitive disabilities (including Down syndrome) who were interviewed had either no friends or only one friend (Snowdon 2012, 6). The research showed that Canadian communities do a lackluster job of including people with cognitive disabilities in social activities. A longitudinal study in the United Kingdom revealed that, as they get older, children with Down syndrome "begin to have fewer social contacts and friends who are not disabled. By the teenage years many are relatively isolated and become increasingly dependent on the family for social interaction" (Cunningham 1996, 89). An expert on the social geography of people with cognitive disabilities, Deborah S. Metzel (2004, 440) notes that, "Research on the friendships and social lives of retarded citizens has revealed that the majority have social contact primarily with people with mental retardation or other disabilities, with staff associated with service provision, and with family." The key to counteracting the "loneliness and social isolation" (440) resulting from this arrangement is the creation of more inclusive communities. At present, people with cognitive disabilities find it difficult to break into the spheres of intimacy that would constitute inclusion—exemplified by meaningful friendships or involvement in informal social groups.

There may be an apparent contradiction here. On the one hand, I have pointed out that people living with Down syndrome report high life satisfaction (Skotko, Levine, and Goldstein 2011c). On the other hand, there is evidence of their social exclusion. But the apparent contradiction might not be substantive. It is possible to enjoy your life and still think that things could be better for you. The reports of high life satisfaction may be a result of the successful inclusion of people with Down syndrome in family life. A few generations ago when children with Down syndrome were sent to institutions, where they lived in abysmal conditions, they were largely excluded from the family realm. The feature that is often missing these days from the lives of people with Down syndrome is successful inclusion in the wider social life of their communities.

These findings about social exclusion touch on one of the fears that parents of children with Down syndrome have about the future of their children. Parents often express their fear of what will happen if their child with Down syndrome outlives them (see, for example, Bérubé 1996, 85). Family life can provide a network of intimates for a person with a disability. Without a network of friends, siblings, or other sources of social association, the death of one's parents may cast one into a forbidding world without the kind of intimate relationships that typically sustain human lives. Adults with cognitive disabilities whose parents have died may have other caregivers who interact with them and look after their well-being. But designated caregivers provided through public social services or through private

---

1 Adrienne Asch and David Wasserman (2005) advance a similar argument that prenatal testing and termination are a rejection of intimacy.

arrangements are inadequate substitutes for meaningful relationships between intimates. We are social beings who need friends and confidants and lovers who value us for who we are in ourselves rather than relating to us out of contractual obligation or professional beneficence.

It is often acknowledged that the relationships we have with people who are closest to us—our family members, our closest friends—result in enhanced ethical responsibilities toward these people. For example, I have a stronger responsibility to look out for the well-being of my child than I do for a stranger. Philosophers have described these enhanced moral duties as the morality of "special relations" (McMahan 2002, 218). For many of us, the most valuable aspects of our lives revolve around these relationships of intimacy—whether they derive from intense friendship, relationships with lovers, or from familial ties. Those who are not given this "special relationship" status are owed the baseline of ethical obligations that prevail among strangers. The danger, of course, is when vulnerable disabled persons become strangers to all because they are intimates of no one.

Social exclusion of people with disabilities—exemplified by the friendlessness of young adults with cognitive disabilities—is one of the effects of bias. These harmful effects coexist with official legislative efforts to diminish discrimination against people with cognitive disabilities, such as the Americans with Disabilities Act. For this reason, it is unconvincing to rebut the disability critique by pointing out that "the rise of prenatal screening has coincided with more progressive attitudes toward the inclusion of people with disabilities, as evidenced in the United States by the passage of the Americans with Disabilities Act" (Steinbock 2000, 121).[2] According to this argument, high rates of selective termination cannot be linked with harmful bias because prenatal testing and selective termination have been developed in societies that have enacted progressive legislation and policies that promote the inclusion of people with disabilities. But informal sources of bias exist beyond the reach of the law; they come to light when we realize, for example, that having a disability can be an intensely lonely experience. It would be impossible to legislate that people with Down syndrome be included in meaningful friendships and other intimate relationships. Nonetheless, these relationships are crucial for the well-being of anyone. We suffer when bias prevents our involvement in such relationships. So inclusion of people with cognitive disabilities into our intimate spheres beyond the family will require changes in common social attitudes of bias that exist outside the jurisdiction of legislation.

References

Asch, A. (2003). Disability Equality and Prenatal Testing: Contradictory or Compatible? *Florida State University Law Review*, 30, 315–342.

Asch, A., and Wasserman, D. (2005). Where Is the Sin in Synecdoche? Prenatal Testing and the Parent-Child Relationship. In D. Wasserman, J. Bickenbach, and R. Wachbroit (Eds.), *Quality of Life and Human Difference: Genetic Testing, Health Care, and Disability* (pp. 172–216). New York: Cambridge University Press.

---

2 Asch (2003) also argues vigorously that legislation like the Americans with Disabilities Act has not led to greater inclusion, even in areas within the reach of the law.

Bérubé, M. (1996). *Life As We Know It: A Father, a Family, and an Exceptional Child.* New York: Pantheon Books.

Bryant, L. D., Green, J., and Hewison, J. (2010). The Role of Attitudes Towards the Targets of Behaviour in Predicting and Informing Prenatal Testing Choices. *Psychology and Health*, 25 (10), 1175–1194.

Choi, H., Van Riper, M., and Thoyre, S. (2012). Decision Making Following a Prenatal Diagnosis of Down Syndrome: An Integrative Review. *Journal of Midwifery & Women's Health, 57*, 156–164.

Coles, S., and Scior, K. (2012). Public Attitudes towards People with Intellectual Disabilities: A Qualitative Comparison of White British and South Asian People. *Journal of Applied Research in Intellectual Disabilities, 25*, 177–188.

Cunningham, C. C. (1996). Families of Children with Down Syndrome. *Down's Syndrome: Research and Practice, 4* (3), 87–95.

Enea-Drapeau, C., Carlier, M., and Huguet, P. (2012). Tracing Subtle Stereotypes of Children with Trisomy21: From Facial-Feature-Based to Implicit Stereotyping. *PLoS One, 7* (4), e34369.

Hein, S., Grumm, M., and Fingerle, M. (2011). Is Contact with People with Disabilities a Guarantee for Positive Implicit and Explicit Attitudes? *European Journal of Special Needs Education, 26* (4), 509–522.

Korenromp, M. J., Page-Christiaens, G. C., van den Bout, J., Mulder, E. J., and Visser, G. H. (2007). Maternal Decision to Terminate Pregnancy after a Diagnosis of Down Syndrome. *American Journal of Obstetrics and Gynecology, 196*, 149.e1–149.e11.

McMahan, J. (2002). *The Ethics of Killing: Problems at the Margins of Life.* New York: Oxford University Press.

Metzel, D. S. (2004). Historical Social Geography. In S. Noll and J. Trent (Eds.), *Mental Retardation in America: A Historical Reader* (pp. 420–444). New York: New York University Press.

Natoli, J. L., Ackerman, D. L., McDermott, S., and Edwards, J. G. (2012). Prenatal Diagnosis of Down Syndrome: A Systematic Review of Termination Rates (1995–2011). *Prenatal Diagnosis, 32*, 142–153.

Nelson, J. L. (2007a). Synecdoche and Stigma. *Cambridge Quarterly of Healthcare Ethics, 16*, 475–478.

Ouellette-Kuntz, H., Burge, P., Brown, H. K., and Arsenault, E. (2010). Public Attitudes Towards Individuals with Intellectual Disabilities As Measured by the Concept of Social Distance. *Journal of Applied Research in Intellectual Disabilities, 23*, 132–142.

Proctor, S. N. (2012). Implicit Bias, Attributions, and Emotions in Decisions about Parents with Intellectual Disabilities by Child Protection Workers. Ph.D. diss,, Pennsylvania State University.

Robey, K. L., Beckley, L., and Kirschner, M. (2006). Implicit Infantilizing Attitudes about Disability. *Journal of Developmental and Physical Disabilities, 18* (4), 441–453.

Skotko, B. G., Levine, S. P., and Goldstein, R. (2011a). Having a Brother or Sister with Down Syndrome: Perspectives from Siblings. *American Journal of Medical Genetics. Part A, 155*, 2348–2359.

Skotko, B. G., Levine, S. P., and Goldstein, R. (2011b). Having a Son or Daughter with Down Syndrome: Perspectives from Mothers and Fathers. *American Journal of Medical Genetics. Part A, 155*, 2335–2347.

Skotko, B. G., Levine, S. P., and Goldstein, R. (2011c). Self-Perceptions from People with Down Syndrome. *American Journal of Medical Genetics. Part A, 155*, 2360–2369.

Snowdon, A. 2012. *The Sandbox Project: Strengthening Communities for Canadian Children with Disabilities.* Report for Sandbox Project's Second Annual Conference, January 19. http://sandboxproject.ca/s/2012-sand-box-conference-mh-snowdon-strengthening-communities-for-canadian-children-with-disabilities.pdf.

Steinbock, B. (2000). Disability, Prenatal Testing, and Selective Abortion. In E. Parens and A. Asch (Eds.), *Prenatal Testing and Disability Rights* (pp. 108–123). Washington, DC: Georgetown University Press.

Wilson, M. C., and Scior, K. (2014). Attitudes Towards Individuals with Disabilities As Measured by the Implicit Association Test: A Literature Review. *Research in Developmental Disabilities, 35,* 294–321.

Wilson, M. C., and Scior, K. (2015). Implicit Attitudes towards People with Intellectual Disabilities: Their Relationship with Explicit Attitudes, Social Distance, Emotions and Contact. *PLoS One.* doi:10.1371/journal.pone.0137902.

Informed Consent or Institutionalized Eugenics?

How the Medical Profession Encourages Abortion of Fetuses with Down Syndrome

Darrin P. Dixon, J.D.

....

There are numerous contributing factors to what some may call a high termination rate of fetuses that have tested positive for Down Syndrome. One major factor is the direct and indirect influences of medical professionals, which include genetic counselors, family physicians and obstetricians and gynecologists. In this article, I support the ethical principle of nondirective counseling and the genetic counselors who seek to achieve nondirectiveness. However, I suggest genetic counselors and many medical professionals have a deference to the use of medical technology and the belief that patients desire the maximum amount of information. This ingrained deference hinders most medical professionals from being neutral and often causes a subtle promotion of prenatal testing and abortion. Overall, increased prenatal testing contributes to the high abortion rate of fetuses diagnosed with Down Syndrome, a lack of genuine informed consent, greater intolerance of people and especially children with disabilities, and less money for research and development of effective treatments. To the extent that women are encouraged to terminate their pregnancies, prenatal testing and abortion of affected fetuses cannot be considered morally justified because the decision lacks genuine informed consent.

Similarly, other medical professionals, such as family physicians, obstetricians and gynecologists, contribute to the problem. Initially, almost all women seek prenatal treatment from a family physician or obstetrician and gynecologist. However, these medical professionals tend to spend significantly less time with patients compared to genetic counselors, which can result in miscommunications. Moreover, these professionals may encourage prenatal testing and the use of "up front" consent forms to reduce legal liability. In addition, these medical professionals typically receive inadequate genetic training, which can result in the misinformation, and most discouraging, undue influence, bias or prejudice against persons with disabilities, which circumvent informed consent. Time constraints, fear of liability, little genetic training and the practice of directiveness can easily result in a negative tone that manifests itself in phrases such as, "I'm sorry," or "Unfortunately, I have some bad news to share" and conversations void of the positive reality that many individuals with Down Syndrome can become semi-independent and with good medical care can live into adulthood. Lastly, both medical professionals and patients and their families may overly rely on genetic technologies, which are far from perfect. The assumption that these technologies are 100% accurate can lead to many injudicious and erroneous choices depending upon the degree of inaccuracy. Yet, medical information is only part of what women and their families use to make their decisions. Their decisions are likely more substantially swayed by societal influences and pressures.

It is important to realize that genetic counselors and other healthcare professionals bring their own values into the prenatal testing process, with patients also adding different and competing values and background knowledge to the process. Furthermore, patients' values reflect a combination of individual perspectives and social norms. While a great deal of variation exists among patients, some trends may be observed. Some women reject prenatal testing because they know they would not have an abortion for moral, religious or personal reasons. Others reject such testing because of the risk of miscarriage. But a vast majority of women at increased risk (those for whom it is medically indicated) of chromosomal or other detectable conditions under the old guidelines choose prenatal screening and/or testing.[1] Several factors contribute to this trend. Just as medical professionals are not neutral about the value of information, neither is our society, which views the gathering of information as a sign of responsible behavior and good decisionmaking.[2] In the context of prenatal testing, patients may believe that getting information about the fetus is not only the right thing to do, but a form of reassurance and a way to get a sense of control over the potentially overwhelming experience of reproduction.[3] This trend contributes to the massive increase in prenatal testing and the need to know whether a child has a disability.

This article does not propose the elimination of prenatal testing. Rather, it proposes that the genetic testing and counseling should not be biased against the birth of children with disabilities. Genetic testing and counseling should not convey directly or indirectly the message that the lives of persons with disabilities are worth less than other lives, or that the only practical alternative is to prevent their existence through abortion.

---

1 Barbara Katz Rothman, The Tentative Pregnancy: How Amniocentesis Changes the Experience of Motherhood 16, 63 (1986). *See also* Gwen Anderson, *Nondirectiveness in Prenatal Genetics: Patients Read Between the Lines*, 6 Nursing Ethics 126, 129 (1999).

2 "In genetics, clinicians and researchers believe that knowledge and genetic science are moral goods." Anderson, *supra* note 1 (noting the public's acceptance of the "moral imperative to know"). Also, observing the medical profession's view that genetic testing will further the well-being of fetus, siblings, parents and society. *Id.*

3 Sonia Mateu Suter, *The Routinization of Prenatal Testing*, 28 Am. J. L. & Med. 233, 246 (2002).

Overall, prenatal testing should be a way for women and their families to reduce the stress and anxiety associated with the unexpected birth of babies with special needs and also a conduit through which women are given information to help them appreciate the value of children with special needs and expand their knowledge of available services and treatment options for such children. However, the practical result of prenatal testing tends to be an increased termination rate of fetuses diagnosed with Down Syndrome or other genetic anomalies. [...].

The Disability Rights Perspective

The premise of the disability rights movement is that persons with disabilities are disadvantaged far more by negative social attitudes than by their disabilities.[4] Disability rights advocates contend that tests like amniocentesis often are performed because a value judgment has been made that there is merit in identifying a fetus who could become a person with a disability. The premise of the expressivist[5] argument is that prenatal testing is morally problematic because it expresses negative or discriminatory attitudes about both impairments and those who carry them.[6] Its central claim is that prenatal tests that expose disabling traits express a hurtful attitude about and send a hurtful message to people who live with those same traits.[7] In the late 1980s, Adrienne Asch, a bioethicist at Wellesley College, put the concern this way: "Do not disparage the lives of existing and future disabled people by trying to screen for and prevent the birth of babies with their characteristics."[8]

Persons or families with disabled children have claimed that a policy that encourages abortion of fetuses with genetic anomalies is a public statement that the lives of people with disabilities are worth less than those of the able-bodied.[9] In addition, such a policy reduces the number of persons with those disabilities, thus reducing their political effectiveness. It also may manipulate couples into carrier and prenatal screening to avoid children with such characteristics.[10] In short, it engenders or

---

4 Asch, *supra* note 57.

5 Expressivists argue that prenatal testing sends the message to people with disabilities that their lives are not worth living. Mahowald, *supra* note 63, at 229. Allen E. Buchanan is noted for developing the "expressivist argument." Allen E. Buchanan, *Choosing Who Will be Disabled: Genetic Intervention and the Morality of Inclusion*, 13 Soc. Phil. & Pol'y 18 (1996).

6 *Id.*

7 Robertson, *supra* note 56.

8 *Id.*

9 Asch, *supra* note 57, at 60–70 ("life with disability can be valuable and valued, and therefore, we must carefully consider the consequences of our disability prevention activities"); Martha A. Field, *Killing "The Handicapped"—Before and After Birth*, 16 Harv. Women's L.J. 79, 123–24 (1993). The article states that the government cannot attempt to eliminate particular disabilities "without disparaging existing persons with that condition by suggesting ... that life with that disability is worse than no life at all."

10 Julia Walsh, *Reproductive Rights and the Human Genome Project*, 4 S. Cal. Rev. L. & Women's Stud. 145, 160–68 (1994) ("arguing that state regulation of prebirth genetics has the potential to infringe upon a woman's right to reproductive freedom").

reinforces public perceptions that people with disabilities should not exist, making intolerance and discrimination toward them more likely.[11]

The message sent, from this perspective, is that a child with the condition would be unacceptable to the prospective parents. This devaluation appears more subtly in the promotion of prenatal genetic testing as helping prospective parents to guarantee that they will have "healthy" children.[12] This rhetoric of good health fails to acknowledge that some traits screened for do not necessarily affect a child's health, although they may impair the child's abilities.[13] A perfectly physically healthy child with Down Syndrome or with deafness is a prime example. Viewed in this light, the appeal to good health, while unobjectionable on its face, may promote eugenic attitudes that individuals with some disabilities are properly excludable, not only from society, but also from existence.[14]

This is a powerful charge, and at the very least should remind us to look more closely at the effects of genetic selection programs on persons with disabilities, and to change the programs if they are in any way harmful, denigratory or disrespectful to people with disabilities. However, the charge is not irrefutable. Surely one can find a particular living situation less preferable than others yet still respect persons in that situation. A policy to prevent accidents that cause paraplegia does not harm existing paraplegics, nor prevent us from supporting programs that make their lives easier. Similarly, a program that enables people to avoid the birth of children with disabilities does not have to denigrate existing persons with those conditions. However, this is not the current state of prenatal genetic testing in this country with regards to Down Syndrome.

While genetic counselors around the world offer prenatal testing as an opportunity to maximize a couple's reproductive choices, disability scholars have recently condemned prenatal testing as typically done with the goal of identifying an affected fetus so that the fetus may be aborted.[15] This message has important, indeed critical, significance for the profession of genetic counseling and should not be overlooked. The disability rights[16] perspective has two central claims: first, prenatal testing is morally problematic; and second, prenatal genetic counseling is driven by misinformation.[17]

---

11 Robertson, *supra* note 56, at 482.

12 Mary A. Crossley, *Choice, Conscience, and Context*, 47 Hastings L.J. 1223, 1230–34 (1996).

13 *Id.*

14 *Id.*

15 Erik Parens & Adrienne Asch, *The Disabilities Rights Critique of Prenatal Genetic Testing: Reflections and Recommendations,* 29 Hastings Center Rep. S1 (1999).

16 This critique, also called "[t]he disabilit[ies] rights critique" is well-developed by Erik Parens & Adrienne Asch, *The Disability Rights Critique of Prenatal Genetic Testing: Reflections and Recommendations, in* Prenatal Testing and Disability Rights 3 (Erik Parens & Adrienne Asch, eds. 2000).

17 *Id.*

However, advocates of prenatal testing argue the widespread use of prenatal testing enables parents to prepare emotionally and financially for the special needs of a child with Down Syndrome.[18]

Another concern is that prenatal genetic testing encourages reductivism.[19] Here, the concern is that using prenatal testing for trait selection (or deselection) purposes will encourage the identification of a specific child with his selected trait(s) or, more generally, the identification of all persons with their selectable traits.[20] The identification of individuals primarily with a single, physical trait rather than with their personhood is precisely opposed to disability rights advocates' efforts to promote "people first" language in describing persons with disabilities.[21] The possibility of reductivism appears particularly troublesome for two reasons. First, it threatens the loss of an intangible aspect of how we view our fellow humans.[22] Our very respect for the dignity of the individual seems premised on our understanding that each individual is greater than the sum of his or her parts.[23] If, by contrast, we were to view our children or the persons with whom we interact in society as simply a combination of traits, then persons with similar traits would begin to appear largely interchangeable, and we would lose an important sense of the humanity and individuality of persons.[24]

Second, reducing our understanding of individuals to the sum of their traits also threatens to create new, and exacerbate existing, biases for social division.[25] Many of the seemingly intractable social divisions of our day are traceable, at least in part, to social groups focusing on one "part" of

---

18 Tracey Hotchner, Pregnancy and Childbirth: The Complete Guide for a New Life 26–35 (1984). *Cf.*, Dorothy C. Wertz, et al., WHO Human Genetics Programme, Review of Ethical Issues in Medical Genetics 62 (2003) (setting forth policy recommendations by health care professionals, including that "prenatal diagnosis can be used to prepare for the birth of a child with a disability instead of making a decision about abortion"); Brian G. Skotko, *Prenatally Diagnosed Down Syndrome: Mothers Who Continued Their Pregnancies Evaluate Their Health Care Providers*, 192 Am. J. Obstetrics & Gynecology 670, 675–76 (2005). (Consequently, health care providers should appreciate that many women consent to prenatal testing with ambivalence or no intent whatsoever to abort.) Driven by a desire to know, many parents will undergo genetic screening even if their only option is to deliver their child: In examining women at risk for having children with sickle cell disease, one investigator who interviewed thirty women found that the majority of them would want prenatal diagnosis even though only one quarter would abort an affected fetus. Another group of researchers who looked not at what women said but what they did found that twenty-two pregnant women who were at risk for having a child with sickle cell anemia, fourteen had amniocentesis, and of the four fetuses found to be affected, three were aborted. Ellen Wright Clayton, *Screening and Treatment of Newborns*, 29 Hous. L. Rev. 85, 116 (1992).

19 Reductivism means the "use of the fewest and barest essentials or elements." Answers.com, *available at* http://www.answers.com/topic/reductivism.

20 Susan M. Wolf, *Beyond "Genetic Discrimination": Toward the Broader Harm of Geneticism*, 23 J.L. Med. & Ethics 345, 347 (1995).

21 "The 'people first' approach prefers describing a person who has a disabling condition as a "person with a disability," rather than a 'disabled person.'" *See* Allan H. Macurdy, *Disability Ideology and the Law School Curriculum*, 4 B.U. Pub. Int. L.J. 443, 443 n.1 (1995).

22 Crossley, *supra* note 80.

23 *Id.*

24 *Id.*

25 Wolf, *supra* note 88, at 347–49.

individuals (for example, their race, ethnicity, religion, or sexual orientation), rather than on their humanity.[26] It is easy to hate and chastise a label; it is more difficult to hate an individual when one views that individual as being a bundle of humanity—with joys, fears, dreams, concerns, vulnerabilities, and strengths.[27] By encouraging us to conceptually break down persons into traits, prenatal genetic testing threatens to reinforce our existing and destructive reductivist tendencies.[28]

Many of the problems we have regarding the normalcy of children center around the fact that we, in the United States, live in a celebrity oriented and visually oriented culture.[29] As long as the status quo remains, the abortion rate will always be high. A common exercise in genetic counseling classes involves asking students whether or not they would choose the traits of their child if they had the option. Furthermore, if they did have the option, would they choose for their child to be taller or shorter, lower or higher body fat, attractive or unattractive features, athletically gifted or intellectually talented, or neither and so on.[30] Almost always students say they would choose their child's traits and that they would choose the more socially advantageous traits.[31] This shows the students value physical appearance, intellectual ability and cosmetic attributes. The root of the problem is that we as a society value some things more than others. More likely than not, children with Down Syndrome do not tend to be thought of as having the above attributes that appear to be desired by parents and the vast majority of society.[32] This problem will not be solved until we as a society accept people who are differently abled and demonstrate that acceptance in our choices and values. [...].

Non-Directiveness: The Unattainable Ideal

A major tenant of genetic counseling is non-directiveness. Genetic counselors are taught to be: educational, nondirective, unconditional, and supportive.[33] The Code of Ethics states that counselors should strive to: (1) seek out and acquire all relevant information required for any given situation; (2) continue their education and training; (3) remain aware of current standards of practice; and (4) recognize the limits of their own knowledge, expertise, and therefore competence in any given situation.[34] With regard to their relationship with their clients, counselors should strive to: "(1) enable

---

26 *Id.*

27 *Id.*

28 *Id.*

29 Gettig, *supra* note 9.

30 *Id.*

31 *Id.*

32 Mahowald, *supra* note 63, at 233.

33 Nat'l Soc'y of Genetic Counselors, *A Voice, a Resource and an Educational Environment for the Genetic Counseling Profession* (1991), distributed at the 1997 Annual Genetics Conference, in Baltimore, MD, American Society of Genetics Counselors, *available at* http://www.springerlink.com/content/t7366q436n21476p/.

34 *Id.*

their clients to make informed decisions, free of coercion, by providing or illuminating the necessary facts and clarifying the alternatives and anticipated consequences, and (2) refer clients to other competent professionals when they are unable to support the client."[35]

The hallmark of genetic counseling is nondirectiveness.[36] Nondirective, or client-centered, counseling is the process of skillfully listening to a client, encouraging the person to explain his or her concerns, helping the client to understand the relevant issues, and determine a course of action.[37] This type of counseling is "client centered" because it focuses on the client, rather than on the counselor. The counselor primarily listens to and tries to help the client discover and follow improved courses of action.[38] They especially "listen between the lines" to learn the full meaning of their client's feelings.[39] They look for assumptions underlying the counselee's statements and for the events the counselee may, at first, have avoided talking about. Counselors often say a person's feelings may be likened to an iceberg.[40] The feelings and emotions expressed by the patient may be only the "tip" of the iceberg. Underlying these expressed feelings and emotions lay the ultimate dilemma to be faced by the patient, which the patient is almost always reluctant to reveal.

Despite their commitment to nondirectiveness, genetic counselors may subtly promote prenatal testing because of values they hold dear.[41] As stated earlier, medical professionals are not neutral about the value of information, and the view that gathering of information as a sign of responsible behavior and good decision–making.[42] Moreover, genetic counselors and patients tend to believe that getting information about the fetus is not only the right thing to do, but a form of reassurance and a way to get a sense of control over the potentially overwhelming experience of reproduction.[43] Also, some popular books have linked the notion of good parents with prenatal testing.[44] Professor Suter has suggested that more genetic counselors hold stronger views about the "rightness" or "wrongness" of abortion for genetic anomalies and its effect on the overall makeup of society than do most physicians.[45]

---

35 Integrated Publishing, *Types of Counseling* (July 1, 2007), *available at* http://www.tpub.com/content/advancement/14144/css/14144_105.htm.

36 *Id.*

37 *Id.*

38 *Id.*

39 *Id.*

40 Suter, *supra* note 3, at 246.

41 Anderson, *supra* note 1 ("In genetics, clinicians and researchers believe that knowledge and genetic science are moral goods" (noting the public's acceptance of the "moral imperative to know.")).

42 Suter, *supra* note 3.

43 *Id.* at 247.

44 Suter, *supra* note 3, at 246; Seymour Kessler, *The Psychological Paradigm Shift in Genetic Counseling*, 27 Soc. Biology 167, 168, 176 (1980). This article discusses the shift from the eugenics paradigm, which focused on managing human heredity, to the current paradigm of psychologic medicine, which focuses on helping patients resolve problems and make decisions. Sonia M. Suter, *A Fresh Look at Nondirectiveness in Genetic Counseling* (Jan. 1, 2000) (unpublished manuscript, on file with author).

45 Gettig, *supra* note 9; Balkite, *supra* note 7; Clayton, *supra* note 20.

Most genetic counselors attempt to achieve the goal of nondirectiveness yet there are countless stories of those who do not. As a result, genetic counselors are not without fault. Although they are supposed to be nondirective, many people simply cannot make a decision. Professor Getting, among other genetic counselors, has experienced women continually asking questions such as, "What would you do in my situation?"[46]

How directive is it to talk patients through the decision-making process? "Talking through" means discussing the things the genetic counselor would consider important and that others have considered important, such as: How stable is the patient's marriage? Is the patient financially capable of raising this child? Is the patient's job flexible enough to handle a child with special needs? Should the patient stay at home instead? Will the patient psychologically and physically be able to handle the special needs of this child? What would it be like to raise this child? How would the child affect the patient's other children? What is the patient prepared to do and not do for the special needs, both physical and psychological, of this child?[47]

Is the above directive or non-directive? It appears the questions asked show value judgments and concern on the part of the genetic counselor. Many genetic counselors believe they can achieve non-directiveness through consciously being aware of their word choice, body language, and maintaining an awareness of the gravity of the information being shared.[48] Furthermore, genetic counselors argue that being nondirective does not mean that you cannot give directions to people, and distributing information on genetic screening to people is not directive because the patient can chose whether or not to follow those guidelines.[49] Although much discussion has occurred regarding nondirectiveness, there does not appear to be any exact guidelines regarding what is and what is not nondirective.[50] The end result is although most genetic counselors may consciously attempt to be nondirective by remaining nonjudgmental and refraining from imposing their beliefs and values, the very words they use and questions they raise may subtly influence the patient's decision more than they realize.

Nondirectiveness, as described in the literature,[51] may oversimplify how counselors understand the counseling experience and it may ignore the very real possibility that genetic counselors do not share a uniform understanding of nondirectiveness.[52] Indeed, no good empirical data exists regarding what nondirectiveness really means to genetic counselors.[53] Moreover, the traditional account of

---

46 *Id.*

47 *Id.*

48 *Id.*

49 *Id.*

50 Sonia M. Suter, *Genetics and the Law: The Ethical, Legal, and Social Implications of Genetic Technology and Biomedical Ethics: Sex Selection, Nondirectiveness, and Equality*, 3 U. Chi. L. Sch. Roundtable 473, n.34 (1996).

51 *Id.*

52 *Id.*

53 *Id.* It is beyond the scope of this article to explore those problems, but they should be mentioned. Other authors have discussed this issue. *See* Suter, *supra* at 239.

nondirectiveness tends to describe a process that is potentially incoherent or inconsistent in some respects.[54]

While striving to achieve nondirectiveness is the correct approach, it is impossible to advance one moral viewpoint over another (i.e., deference to technology, knowledge and the like) and be neutral toward all moral viewpoints. Professor Christy A. Rentmeester, Ph.D.[55] said it this way, "Despite the best efforts of a counselor to convey "value neutral" facts, risk assessment by the counselee and family is done according to normative analysis, experience with illness, and definitions of health. Each of these factors must be known by the genetic counselor in order to relate those facts which she acknowledges as relevant to the decisions that will be made by those people seeking the genetic information. In the expression of genetic risks, the authority of medical language impacts a person's understanding of epidemiological data."[56] Moreover, the strong values genetic counselors place on knowledge, information, and technology likely reinforces the public's acceptance and expectation of prenatal testing.[57] Thus, even a genetic counselor's best effort to be nondirective has some elements of directiveness.

---

54 William A. Galston, Liberal Purposes: Goods, Virtues, and Diversity in the Liberal State 80–81 (1991).

55 Assistant Professor, Creighton University Medical Center, Center for Health Policy and Ethics.

56 Christy A. Rentmeester, *Value Neutrality in Genetic Counseling: An Unattained Ideal,* 4 Med., Health Care & Phil. 47–51 (2001), *available at* http://www.springerlink.com/content/wu0r2vw-23w42q5p7/.

57 Gettig, *supra* note 9.

DISCUSSION QUESTIONS

1. How do genes determine who we are? What other factors exist?

2. Can we make choices that go against what may be genetically determined?

3. If there were traits (sexual orientation, likelihood of cancer, eye color, height, ability to hear) that you could change about your future child, would you change them? Which would you change? Why those?

4. How would choosing certain traits expand racism, sexism, homophobia, and transphobia?

5. Should parents have a child for the purpose of saving the life of a child already born?

6. What is required to make a truly autonomous decision?

7. How do we balance the freedom to make our own choices with the impact those choices can have on society?

8. How do societal preferences impact decision-making?

Recommended Cases

Molly Nash and Baby Adam—Nash was born in 1994 with Fanconi anemia, a rare genetic condition in which the body cannot make healthy bone marrow. Sufferers rarely reach adulthood. Her parents went to a treatment center, where embryos were produced through IVF and then genetically tested to ensure the absence of Fanconi anemia and immunologically tested to ensure a tissue match with Molly. The one embryo that met both criteria (of the 14 or so created by IVF) was transferred into Mrs. Nash to create a possible donor sibling for Molly. The Nash family had a very long, drawn out but eventually successful treatment, resulting in the birth of baby Adam in 2000. Blood from his umbilical cord was collected at the time of his birth, and stem cells from it have been successfully used as a bone marrow graft for Molly.

DEX for CAH—The intervention of prenatal dexamethasone (DEX) is aimed at causing congenital adrenal hyperplasia (CAH)-affected female fetuses. CAH often causes ambiguous genitalia in female fetal development. This off-label use is to cause these fetuses "to develop in a more female-typical fashion than they otherwise might. Androgens contribute to sex differentiation, including in the brain and genitals; relatively low prenatal levels ordinarily result in a more female-typical development; relatively high levels usually result in male-typical development."[1]

---

1 Dreger, A., E. K. Feder, and A. Tamar-Mattis. "Prenatal Dexamethasone for Congenital Adrenal Hyperplasia." *Bioethical Inquiry* vol. 9 (3): 277–294 https://doi.org/10.1007/s11673-012-9384-9

UNIT IV

Reproduction

··

Unit IV Introduction

The most private matters often become the most politicized. The Hippocratic oath forbode pessary for abortion. In the early 1800s, abortion was legal in the United States so long as it was before "quickening," the start of fetal movement. As more restrictive laws regarding abortion were enacted, the courts were forced to intervene, most notably in *Roe v. Wade* (1973), a decision which made abortion available in all states, declaring the choice of abortion was a matter of privacy. While open debates regarding abortion are often counterproductive, examining the positions is notable. Each side has a pro: pro-choice and pro-life. In a typical debate setting, the opposing sides are for and against. In regard to termination of pregnancy, each side is for a position and casts the other as their opposite.

Reproduction ethics are involved in processes beyond abortion. Prenatal genetic screenings are expensive, as are *in vitro* fertilization and surrogacy. Trying to create a just solution is difficult, as limited access can come from cost, geographic distance, as well as legality. Understanding what harms are caused and which individuals are helped will be determined by our understanding of personhood. Autonomy will be challenged. All of this will be examined in regard to one of our most private matters.

"Ethical Issues in Reproduction" integrates politics and ethics across a variety of issues in reproduction, including termination of pregnancy, surrogacy, and prenatal testing, among others. An ethical framework alone cannot offer access or safety. The paper focuses on the intersection of ethics and politics as a way of protecting principles of autonomy and justice.

We are introduced to the notion of "designer babies" in "Ethical Challenges and the New Technologies of Reproduction." While techniques for creating designer babies are in their very introductory stages, there are some technologies that make us more able than ever to make choices regarding the children people can have. Some of these include attempting to have a savior sibling or choosing to have a child who is deaf.

Designing Babies

Brenda Almond

....

Within the medical field, there are other uses of the new technologies that are often approached on a case-by-case basis. Their object is not normally to fulfill purely social choices, but to help people avoid the trauma some inherited conditions can bring within families. These other choices are made possible by using the techniques of fertility treatment, in particular IVF (in vitro fertilisation), to create a number of embryos and select among them in order to base a pregnancy on a more secure basis or to avoid risking the birth of a child with a serious inherited condition. This technique, known as PGD, can enable parents and their medical advisers to select an embryo of a particular sex, to select for or against a disability, or even to choose an embryo as a potential tissue or bone marrow donor for another family member.

These various possibilities raise a number of ethical questions. But first, it is important to correct some common misunderstandings of the term 'designer babies'. This is often thought to mean that the embryo's genetic structure has been altered to create a 'designer baby' in a positive sense, or that prospective parents have chosen children for characteristics like intelligence, sporting prowess or good looks. But these are complex characteristics involving a number of genes and if such a scenario is ever to be possible, it is a long way in the future. What is possible now is to use the new technologies of assisted reproduction to avoid an undesired medical outcome for a child, usually a genetic condition that can be inherited within a family. Since some well-known genetic conditions are sex-linked, this can often be done simply by selecting an embryo of the unaffected sex. Whether people should be allowed to select for sex for other than medical reasons is a more controversial issue. Some regard state interference to prevent this as an infringement of personal liberty. Again, however, there are broader aspects to be considered. For example, in groups where there is a cultural preference for males, it could place pressure on women to use this difficult and possibly hazardous route to pregnancy. On the other hand, many people would be sympathetic to a couple's wish to 'balance' their family in the sense of having children of both sexes.

However, even if limited to avoiding a serious medical risk, PGD still arouses controversy. In particular, some of those representing people with disabilities believe that if you are trying to eliminate known health problems, this implies that people with those problems are less worthy of respect than other people. They would prefer to see a more positive approach to those conditions and more help for families coping with them.

Disability advocates could, however, take this further and seek to use the technology to select for a condition that is usually regarded as a handicap. Finally, the technology can be used to bring a child into the world for someone else's good or well-being. Two examples can be used to illustrate these last two possibilities, both of which are controversial in that they seem to involve using the technology for ends that are not primarily directed to considering the good of the future child at all, but to some other end.

Choosing a Deaf Child

People who are deaf often share a friendly and supportive community life, hence some deaf people would prefer to have children who are also deaf and so will be able to become members of their own community. In a much-discussed case in the US, this wish led to a deaf woman successfully seeking a deaf sperm donor in the hope of having a child who would also be deaf. (In this case, a sperm donor was sought because the would-be mother was a lesbian with a female partner.) The ethical challenge here is that, no matter how happy the resulting child might appear to be later in life, it seems to have been deliberately chosen to have a life that would lack something that other human beings take for granted and which most people regard as of very high value.

Choosing a Saviour Sibling

The background to the second example is that children suffering from certain rare inherited diseases may be helped by a blood or bone marrow transfusion from a suitable donor, ideally a sibling. Parents may understandably hope that one of their existing children could provide a tissue match and if this is not the case, they may decide to have another child in the hope that the new child will be compatible as a donor for the sick child. Science can now assist this choice and the procedure involved is relatively straightforward, given the widespread use of assisted reproduction. It involves producing a number of embryos, which can then be examined in vitro to see if there is an embryo free of the condition that could become a child who is an ideal tissue match for the existing child.

Compassionate though this might seem, the question is whether the donor child is being fairly treated. Both the child's welfare and the child's rights are involved, and it is doubtful whether either can be adequately protected in these circumstances. The hope is that all that will be needed is the newborn baby's cord blood. However, if the cord blood donation fails, the new child is destined for a possible lifetime of pressure to donate whatever its ailing older sibling might need in the future. This could include not only repeated donations of bone marrow, but even non-replaceable organs. Sometimes an imaginative work of fiction can make an ethical point more clearly than an academic argument and in *My sister's keeper* Jodi Picoult paints a compelling portrait of the dilemmas that could be faced by a child conceived in this way (Picoult, 2004). The child in this novel seeks out a lawyer to help her challenge her parents' right to make medical decisions on her behalf. The ethical challenge here is that, no matter how excellent their situation in other respects, the 'saviour sibling' is a child who has been created in order to be used for a purpose that goes beyond that child's own interests.

References

Picoult, J. (2004) *My sister's keeper*, London, Hodder.

Ethical Issues in Reproduction

....

The Ethics and Politics of Reproductive Choice

To no one's surprise, most of the pressing ethical issues that arise in the clinical setting concern the beginning and the end of life. What may be surprising is that, as we have become more knowledgeable, more skilled, and more technologically savvy, we have also become less rather than more certain about when and how life begins and ends. Solving those mysteries is certainly beyond the scope of this book but this chapter and chapter 9 present ethical issues that your committee may be asked to address.

Reproductive liberty, the freedom to have children, to determine how many to have, and when and with whom to have them is one of the most highly valued liberties in our society. Because reproductive decisions are thought to be among the most personal and deeply meaningful ones that men and women make in their lives, interference with reproductive choice cannot be taken lightly. Because reproductive choice is a fundamental liberty, governmental interference with it can only be morally justified for compelling state reasons and, even then, the interference must impose the least restriction on reproductive choice, consistent with the aims of the intervention.

Given the intimate nature of reproductive issues, one might reasonably think that decisions about them would be confined to the individuals, their partners, and their physicians. One would be wrong. Paradoxically, these most personal issues are also those on which others—family, friends, legislators, religious leaders, politicians, special interest groups, and other complete strangers—feel not only the right but also the obligation to intervene or at least comment. Moreover, their opinions and interventions are highly variable, including those based on proven science and medical evidence, those rooted in theology or law, those intended to further a particular ideological agenda, and those informed by personal religious or philosophical convictions.

When these matters come to the attention of your ethics committee, it is important to be mindful of their polarizing potential and the need to approach discussions with respect and restraint. Patients, families, and caregivers all bring with them their values, beliefs, and personal histories. Your role is to provide the ethics lens through which the parties can view the medical issues and clinical decisions before them, and to keep the focus on those issues. Your concerns are promoting the right of the capable patient to make autonomous, informed, and voluntary decisions; protecting the right of caregivers to decline participation in any activity that violates their religious or philosophical convictions; and preserving the privacy and dignity of the deliberations. Mediating these delicate issues requires interpersonal skill, cultural sensitivity, and focus, and this is precisely what the institutional ethics voice can do most effectively.

While reproductive liberty is greatly valued, there is controversy concerning its ethical contours. New technologies expand reproductive options for men and women, but they also raise questions about their proper use. Should in vitro fertilization (IVF) be used to assist post-menopausal women to become pregnant? Should preimplantation genetic diagnosis be used for the purpose of sex selection? Should gestational surrogates be paid? Should there be a market in sperm and eggs? Reproductive choice, after all, does not just involve the rights and interests of the woman who is contemplating becoming pregnant. They also concern the medical professionals who assist her, if she seeks or needs assistance; the man, woman, and sometimes additional individuals who donate gametes to make the pregnancy possible; the future child; and society at large, whose norms regarding reproduction may be adversely affected by the reproductive choices that are permitted within it. There are ethical issues in relation to each of these parties, and how they are addressed determines what the right to reproductive choice is actually a right to do.

Reproductive liberty is not, of course, the entire story about the ethics of reproduction. There is also an ethics of reproductive responsibility, and arguments for limiting reproductive liberty often come from this quarter. Having children knowing that one cannot adequately care for them; having a child with a known high risk of a serious genetic disease; or exposing the fetus to alcohol, drugs, or other harms raises the question of whether parents are acting responsibly in how or whether they reproduce.

Parents may be charged with acting irresponsibly if they knowingly bring a child into the world who will suffer from a serious disease or disabling medical condition. Against this, some philosophers have argued that this child is not harmed by being born in this condition since he is not made worse off than he would otherwise have been; he otherwise would not have been at all (Parfit 1984). Others have proposed that, even if this child is not harmed by being born, he is born in a harmed condition, and parents may have a duty to prevent the resulting suffering (Brock 1995).

Concerns about reproductive responsibility can be taken too far when, for example, a woman's reproductive freedom is constrained in the misguided belief that she is not fit to be a parent. To be sure, parental fitness is a legitimate concern but one that is properly based on a person's behavior that has been *demonstrated* to constitute child abuse or neglect. It is certainly not within the scope of practice for health care professionals to *predict* parental fitness and unilaterally restrict the reproductive liberty of those they find wanting. The early twentieth-century witnessed a shameful period when people who were considered socially undesirable—mainly women, people of color, and those who were or had been in prison or psychiatric asylums—were involuntarily sterilized for the good of society. Egregious cases, such as *Buck v. Bell*, discussed in part IV, resulted in strict standards and safeguards to prevent actions justified by the notion that the right to reproductive choice should be balanced against social norms and interests (Suter 2011).

Reproduction is also not just an ethical issue in our society, it is a deeply and an inextricably political one, as well. Differences of opinion about the extent of women's legitimate reproductive freedom are intimately tied up with differing views about the proper role of women and the nature of the family, and they derive much of their intensity from this connection. The right to abortion is still vehemently debated and contested, serving as a political litmus test for politicians on different ends of the political spectrum and continuing to divide opinion more than 40 years after it was legalized in *Roe v. Wade*. States pass restrictive abortion laws and punish women for endangering the health and lives of the children they intend to bear. Religious institutions refuse to provide contraceptive health coverage for their employees and refuse to adhere to government mandates requiring them to do so. There is no set of issues in the field of bioethics that is as heavily and passionately politicized as reproduction.

The polarizing debate about reproductive choice, especially abortion, has very direct and practical implications for health care providers. In an increasing number of states, the effort to overturn *Roe v. Wade* has been supplemented with statutes that require or prohibit what health care providers discuss with women who ask for information about pregnancy termination. These statutory maneuvers are described further in the discussion of abortion below. The same state intrusion into the provider-patient dynamic prohibits discussion of certain other topics as well, such as the ownership, storage, and safe use of firearms. Regardless of one's position on controversial issues, such as abortion or gun control, it should be a matter of concern to health care providers and ethics committees when states legislatively control what practitioners discuss with their patients as part of their health care.

Assisted Reproductive Technologies

Henrietta Perkins is a 55-year-old divorced high school teacher who recently married John Franklin, 48, a man she met through an online dating service. Henrietta is close to her two grown children from her first marriage, but she is eager for another with her new partner. She and John have talked about the possibility of having a child of their own and both of them feel that this would enhance their marriage. Henrietta is post-menopausal, however, and to conceive she would need both donor eggs and reproductive assistance through IVF. Henrietta's two grown children are not entirely on board with her plan because they are not sure where the donor eggs would come from and they worry about whether their mother would be able to manage child care as she and the child age. The fertility clinic that Henrietta and John visit has the same concerns and is unsure whether it is ethically and medically appropriate to facilitate their plan. The clinic seeks the advice of the reproductive technologies ethics committee to which it frequently brings difficult cases.

Whose interests should count here and in what priority? What risks are associated with the use of IVF in this case, and how should they be weighted? What role does reproductive freedom play here? Would it be ethically responsible for a physician to assist Henrietta in having a child?

In Vitro Fertilization

According to a 1995 survey conducted by the National Center for Health Statistics, an estimated 2.1 million or 7.1% of married couples with a wife of childbearing age are currently infertile. Infertility is defined in the medical literature as difficulty achieving conception after a specified period of time, usually 12 months, despite unprotected intercourse (National Center for Health Statistics 2011). For many couples, infertility is a source of great frustration, pain, and disappointment. Fortunately, with the advent of techniques of assisted reproductive technologies (ARTs), chiefly IVF and its offshoots, such as intracytoplasmic sperm injection and gamete and zygote intrafallopian transfer, infertile couples who want children may be enabled to conceive and complete a pregnancy.

While IVF is not a panacea for infertility, notable improvements have been made since its first documented use in 1978. A single attempt at IVF costs anywhere from $5,000 to $12,000, and it is not uncommon for couples to attempt more than one cycle since the initial cycle is often unsuccessful. According to statistics from the Centers for Disease Control (2012), a woman who uses donor eggs or non-frozen embryos has a 56% chance of having a live birth and a 37% chance of a singleton live birth, increased from 2002, when even the most successful clinics working with the healthiest couples reported success rates only in the mid-30% range. The use of IVF has expanded beyond infertile heterosexual couples in which the women are of childbearing age and now includes men and women without partners, gay and lesbian couples, and post-menopausal women.

Despite improvements in the rate of successful pregnancies, assisted reproduction still raises health concerns. There is a greater chance of multiple births with IVF (and also with fertility drugs) when, as is often the case, multiple embryos are implanted to increase the probability of achieving a pregnancy. Multiple births often lead to prematurity and its attendant, often serious, medical problems for both mother and babies. Miscarriage and premature delivery are common complications of multiple pregnancy, the risk rising exponentially with the number of fetuses (Dickens and Cook 2008). Reports of high-order multiple pregnancies, such as the case of the woman delivering eight babies in her second IVF-created multiple pregnancy (Saul 2009) have increased public awareness of these issues. Heightened scrutiny of the risks posed by multiple pregnancies has raised the question of whether physicians performing IVF should be required to limit the number of embryos implanted, a practice already in place in other countries (Saul 2009).

In addition to health concerns, ARTs pose several ethical questions. In the simple case of IVF, eggs are taken from the woman and sperm from the man since there is either male or female infertility or both; the eggs are fertilized in a laboratory dish and implanted in the woman's uterus 48 to 72 hours after fertilization.

Like abortion, tubal ligation, and vasectomy, ARTs have religious implications. For example, these techniques conflict with the tenets of Catholic doctrine, which forbids any intervention that inhibits conception or substitutes for the conjugal act. Thus, IVF as described in the scenario above, as well as virtually every ART, including artificial insemination by the husband, is opposed by the Catholic Church (Congregation for the Doctrine of the Faith 1987; *Ethical and Religious Directives for Catholic Health Care* 2009). Absent a blanket prohibition on the use of ARTs, a number of ethical issues must be addressed, including respect for the autonomy and religious convictions of patients, partners, and health care providers, and the obligations of providers to promote the well-being of patients, enhance the likelihood of a healthy pregnancy and successful birth, and avoid actions likely to cause harm to mother or fetus.

Variations on the simple case raise additional problems. A couple may undergo IVF with their own gametes and the resultant embryo is then placed in the uterus of another woman, known as a gestational carrier. Critics of such arrangements worry that they are exploitative of the carrier, who may be susceptible to financial pressures, and that they commodify gestation, making the child into an object of monetary value. Further, a woman may receive donor eggs from a variety of sources, including her sister or daughter. These sorts of possibilities have led to worries that IVF can create novel forms of familial arrangement and unsettle the future child's sense of identity, which has traditionally been grounded in biologically unambiguous relationships to others in the family. The general concern here is that IVF forces us to rethink notions of parenting and family that most people have simply taken for granted. With IVF, it becomes possible to assign different parenting roles to different individuals: one individual can serve as the gestational parent, others as the genetic parents, and still others as the social or rearing parents. The close bond linking these different functions could become undone and that, by itself, may be unsettling to some.

Other important ethical questions concern who should have access to assisted reproduction and who should make that determination. Access to IVF and other forms of ART is arguably part of the right to reproductive liberty and, given the fundamental importance of this liberty, there are grounds for claiming that it is unjust to bar or limit access simply because the woman or couple lacks the financial means to pay for it. This, however, is what Medicaid, the public insurance program for the poor, does because it views IVF as an elective procedure that it will not cover. Further, whatever rights flow from reproductive freedom are not absolute; they must be balanced against considerations of reproductive responsibility. As with traditional reproduction, the way ART is utilized is not always considered responsible and, for this reason, unrestricted access to ART may not always be ethically defensible. Consider, for example, a case in which an HIV-positive woman seeks reproductive assistance, but refuses to take any measures to lower the risk of transmission to her offspring.

An additional moral issue that does not arise with traditional reproduction is the rights and responsibilities of the medical professionals who are enlisted to assist individuals and couples to become parents. In addition to the medical factors considered in assessing people requesting IVF, practitioners' attitudes about same-sex unions, racial or ethnic background, and, most recently, gender identity of potential parents may test their ability to make clinical assessments uninfluenced by personal bias (Richards 2014). The medical professional is an independent moral agent who must make her own moral judgments about the appropriateness of providing ART, based on her best understanding of the professional and ethical norms that apply to the case at hand, and these may diverge from those of her patients. Your ethics committee may be asked to weigh in on these matters in consultation or policy review, and your familiarity with these complex matters and their ethical implications will be important.

There are other concerns of a policy nature related to IVF. Currently, the infertility industry in the United States, unlike that in several European countries, is largely unregulated, leaving individual clinics free to charge whatever they want for their services, implant as many embryos in their patients as they see fit, set how much they will pay for donor eggs, and determine what will happen to frozen embryos that are not used by the couple for whom they were created. Weighty and controversial ethical issues, such as whether gametes should be treated as a commodity or whether there should be any constraints on the disposition of unused embryos, are raised by these practices, and it is unsound public policy to leave them entirely up to individual clinics. They argue for greater public deliberation about the merits of these different practices and, informed by this deliberation, a stronger role for the state in regulating them.

The case of Henrietta and John raises a number of concerns, all of which cluster around the issue of the age of the mother. Pregnancy over the age of 50 is associated with a higher risk of medical complications, including hypertension, gestational diabetes, and preeclampsia. There is also an increased risk of genetic defects in the fetus. An additional concern is the respective ages of mother and child; as Henrietta ages, she may have difficulty taking care of an active adolescent when the child is 15 and she is 70. Yet, while the well-established clinical and genetic risks of pregnancy may be within the clinical purview of the benefit-burden-risk assessment and, when the risks are unacceptably high, may justify refusing to perform assisted reproduction, predicting and assessing parental fitness as the woman ages is a more difficult and controversial matter. Finally, there are the important factors related to egg donation, including the identity of the donor and how donation was solicited and reimbursed, discussed below.

As a member of the ethics committee, you should point out that, although it may be "unusual" or "nontraditional" for post-menopausal women to bear children, that is not the same as "unnatural" or "unethical." Instead, your ethical analysis should focus on the rights and health risks of the mother, the medical and psychological welfare of the future child, and the circumstances under which donor eggs are obtained.

Pre-Implantation Genetic Diagnosis

Sybil and Graham Wray, both in their early thirties, are the proud parents of a little girl, Jill, age 5, who suffers from a deadly genetic disease that causes bone marrow failure, eventually resulting in leukemia.

Her best chance for survival is a bone marrow transplant from a perfectly matched sibling donor. Sybil and Graham had considered having another child, but decided against it because of a one-in-four chance the child would have the same disease as Jill. But then they heard about preimplantation genetic diagnosis (PGD), a technique that would enable them to screen out embryos affected with the genetic disease before implantation. Because this technology holds out the prospect of life-saving treatment for Jill, as well as the opportunity to give her a sibling who will not develop the same lethal disease, Sybil and Graham have decided to try PGD.

Whose interests are being served here? Is one person being created to be "used" without his consent to promote the interests of another? Do parents have the right to do what the Wrays are proposing?

Pre-implantation genetic diagnosis has been available since 1990 in conjunction with IVF. One cell (blastomere) is removed from a cleaving embryo ex utero and tested for a particular genetic or chromosomal abnormality. Embryos with these abnormalities can then be discarded, avoiding the birth of a child with a disabling condition, and the embryos without the chromosomal or genetic defect can be transferred to the woman's uterus. New uses of PGD include detecting mutations for susceptibility to cancer and for late-onset disorders, such as Alzheimer's disease (Robertson 2003).

Some uses of PGD are relatively ethically uncontroversial, such as detecting lethal or serious genetic conditions, including Tay-Sachs disease or autosomal recessive polycystic kidney disease (ARPKD), which will profoundly afflict the child from birth. Some commentators have even argued that, in these cases, potential parents have, not just the right, but also an ethical duty, to use PGD to prevent such conditions (Malek and Daar 2012). Somewhat more controversial is using PGD to detect genetic mutations, such as BRCA1&2 genes, which signal significantly heightened risks of cancer and other late-onset disorders, such as Huntington or Alzheimer's diseases. The concerns are that, while early detection of the BRCA genes enables either a decision about terminating the pregnancy or alertness to the need for close monitoring and possible prophylactic intervention for the affected child, detection of susceptibility to Huntington or Alzheimer's risks preventing the birth of a child who may have a healthy and problem-free life for many years.

More controversial still is PGD for the purpose of sex selection. The use of PGD to select the sex of one's offspring is susceptible to the charge of sexism, which is a compelling reason to oppose it. However, parents may also desire to use PGD for sex selection in order to achieve gender balance in their family, which may not be sexist. Nevertheless, the question remains whether achieving family balance is a strong enough moral reason to discard embryos (Ethics Committee of the American Society of Reproductive Medicine 1999).

More speculatively, PGD might be used to select embryos with certain desirable qualities and nonmedical traits, such as perfect pitch or superior intelligence, and to discard those that fall short. These nonmedical uses of PGD, often referred to as creating "designer babies," are especially worrisome because they invoke the reasoning of the early twentieth-century eugenics movement and permit a significantly increased degree of arbitrary reproductive control that may be detrimental to the resulting offspring and, ultimately, to society.

To return to the case of Sybil and Graham Wray, they want to have another child who, in addition to completing their desired two-children family, can serve as a bone marrow donor for Jill, and they want to use PGD for this purpose. Some may find this morally objectionable because it smacks of using a person as a mere means to serve the interests of others. But the case is more complicated than that, or at least it would be, if the Wrays do not consider the second child as a "mere" means. There is, after all, no indication that the Wrays would not love their second child as much as they love Jill or that they would neglect her once she has served her purpose of aiding Jill. As a member of the ethics committee, you could helpfully point out that individuals often have children to serve a variety of purposes of their own, some legitimate and some not, and you could help clarify the differences between those cases in which this is morally acceptable and those in which it is not.

Gamete Donation

Contributing to worries about the commodification of childbearing is the current practice of buying and selling donor eggs on the open market (Daniels and Heidt-Forsythe 2012). Egg donation has become a significant part of the multi-billion-dollar-a year infertility industry. Men who donate sperm receive a small payment; the going rate for eggs is much higher. It is common knowledge at colleges and universities that young female students sometimes decide to donate eggs, assuming the associated medical risks, as a source of income. Infertile couples now offer large sums of money for donor eggs from women who meet certain specifications, such as educational attainment and athletic ability. Payment of some sort for donor eggs seems acceptable and only fair since the process of donating eggs is both burdensome and potentially dangerous. However, if payment is too high, needy donors may be led to discount the risks of donation, making their decision to sell their eggs less than fully voluntary. In addition, infertility clinics may, in their eagerness to obtain high-quality eggs and increase their profits, fail to fully reveal all the risks and possible costs of donation to potential donors, rendering their decision to "donate" less than fully informed. While these issues are or should be addressed during extensive counseling prior to the donation, your ethics committee is encouraged to devote time to their examination.

Posthumous Sperm Retrieval

The chair of your ethics committee has received the following request from the nurse manager in the surgical ICU (SICU): "Okay, I'm in way over my head on this one and I need an ethics consult ASAP. We have a patient who's just been declared brain dead and his wife is asking to retrieve his sperm. Oh, and just to add interest, his parents hate her and they want the sperm."

Mr. O'Reilly, a 34-year-old man, had been admitted with massive brain injury following a head-on motor vehicle accident. After completion of the relevant examinations and tests, the several neurologists agreed that he had lost entire brain function and met the criteria for a declaration of brain death. His immediate and extended family was in attendance and, after careful and detailed explanation, everyone accepted the fact that the patient was now clinically and legally dead. That, however, was where concurrence ended.

Two groups of family members asked to speak privately with the ethics consultant. Mrs. O'Reilly explained that she and her late husband had been trying to start a family and, for the past six months, had been working with an infertility specialist. She tearfully described their shared dreams of parenthood, ending with, "All I have of him now are those dreams and I want so much to make them come true. It's what he would want." Her brother said, "He was so looking forward to being a father." The patient's parents and siblings insisted, "That tramp just wants a kid to inherit his money. She would be a terrible mother. If we can find a nice girl to have his baby, we'll have a grandchild to carry his legacy."

Whose interests are at stake here? What obligations are owed to the late patient, his grieving wife, and his distraught parents? What ethical principles should inform analysis of the situation? What is the appropriate role of the ethics consultant?

Let's begin by clarifying what the purpose of ethics consultation in this case is *not*, which includes determining the validity of Mrs. O'Reilly's claim that she and her late husband wanted to have a child, assessing her character or parental fitness, or mediating family conflicts that predate Mr. O'Reilly's accident and death. The contribution of ethics analysis here is to identify the competing interests and obligations, and ensure that the rights of the now-deceased patient are not violated. It is important at the outset to distinguish this situation from surrogate decision making as discussed in chapter 2. Typically, the surrogate is responsible for making care decisions on behalf of a patient whose welfare is the paramount consideration. Here, the decision will not affect the now-deceased patient's health, only the respect for his prior preferences and the interests of his survivors.

Sperm is commonly procured and preserved in situations in which infertility can be foreseen, such as impending chemotherapy. But retrieval of sperm in cases of unexpected coma or death, while often technically feasible, is ethically extremely problematic. Of major importance here is the deceased man's prior preferences about becoming a parent, specifically whether he consented to retrieval of his sperm for the purpose of fathering a child under the current circumstances. In other words, while Mr. O'Reilly may have ardently desired to become a parent and participated in infertility counseling and ART, those preferences and actions are not evidence of his willingness to *father a child after his death*. For this reason, implied consent is very problematic because of the likelihood that emotionally involved third parties may erroneously assume that consent has or would have been given or that consent would not have been given.

If Mr. O'Reilly had explicitly consented to have his sperm retrieved posthumously for the purpose of reproduction, it would be ethical to proceed under specified conditions. Many hospitals, as well as sperm and egg banks, have policies that require written documentation of the man's prior consent; designation of the intended sperm recipient who will be the child's parent; and prohibition of third parties, such as potential grandparents, having control of the sperm. If the request for sperm retrieval is on behalf of a man who is currently incapacitated, about to begin treatment that is fetotoxic, and expected to regain capacity following treatment, some policies stipulate that a court-appointed guardian assume responsibility for the storage and disposition of the sperm until the man is able to take control or he is permanently incapacitated or has died. But if a man has steadfastly refused to have a child while he was alive, it would be ethically wrong to retrieve his sperm after his death, effectively making him a father against his will (Orr and Siegler 2002). Because any requests for sperm retrieval in your institution will likely generate a clinical ethics consultation, your committee is encouraged to address the matter proactively through discussion and policy development.

Surrogacy and Gestational Carriers

Baby M was born to Mary Beth Whitehead, conceived with her own oocytes and sperm from William Stern, the husband of Elizabeth Stern, the intended child-rearing parents. The Sterns and Ms. Whitehead entered into a contract according to which Ms. Whitehead would relinquish her parental rights in favor of Mrs. Stern upon the birth of the child. However, she decided to keep the child, so the Sterns sued to be recognized as the child's legal parents. In an important decision by the New Jersey Superior Court, the contract was ruled invalid according to public policy, Ms. Whitehead was recognized as the child's legal mother, and family court was ordered to determine whether Ms. Whitehead, as mother, and Mr. Stern, as father, should have legal custody of the child, according to the traditional "best interests of the child standard." Ultimately, Mr. Stern was awarded custody of Baby M, Mrs. Stern legally adopted her, and Ms. Whitehead was given visitation rights.

Are there legitimate reasons to prevent people from becoming or hiring surrogates or gestational carriers? Is the practice so dangerous, risky, or contrary to the public good that it should be banned? What ethical values are promoted by permitting the practice?

First a word of clarification. A *gestational carrier* is a woman who carries to term a child conceived with the gametes of a couple to whom she relinquishes the child upon delivery. As such, she is a parent only in the gestational, rather than the genetic, sense. A *surrogate*, by contrast, provides her own oocytes, fertilized with sperm from the man in another couple to whom she relinquishes the child upon delivery. Her contribution is both gestational and genetic. Mary Beth Whitehead was a surrogate.

Surrogacy is far more accepted in the United States than in most countries; however, there is no national consensus on how to deal with it even here. As of 2014, 17 states have laws permitting surrogacy, some with restrictions; 6 states prohibit enforcement of surrogacy contracts; and in 21 states, there is neither a law nor a published case regarding surrogacy (Lewin 2014).

The ethical arguments that support allowing women to serve as surrogates or gestational carriers are rooted in several values: autonomy, beneficence, and reproductive freedom. According to the autonomy argument, women should be free to make decisions about their own bodies, including waiving their parental rights before the birth of children they help conceive or carry. The beneficence argument emphasizes the good that surrogates and gestational carriers can provide couples whose desire for a child with all or part of their genetic makeup has been impeded by the woman's infertility or inability to carry a pregnancy to term. Surrogacy also enhances a woman's reproductive freedom to have a child to whom she is genetically related.

Despite these benefits, there are many reasons to be cautious about surrogacy and gestational carrier arrangements. Relinquishing a child whom one has carried to term can be emotionally traumatic for the carrier, as happened in the Baby M case. An additional concern is that women may agree to serve as surrogates because they see this as an opportunity to improve their economic situations and, as a consequence, the inducement of financial compensation is potentially exploitative. According to 2008 statistics, the mean compensation for a gestational carrier in the United States was approximately $20,000 (Brezina and Zhao 2012). Financial compensation for the delivery of a baby also seems to be tantamount to baby selling, which is both immoral and against public policy. These concerns, while worth taking seriously, can be addressed and should not justify complete prohibition of the practice of surrogacy. The concern about the emotional cost to the surrogate or carrier can be addressed to some extent by pre-conception counseling and requiring a waiting period before the surrogate relinquishes her parental rights; exploitation is less a worry if financial compensation is limited to payment for medical expenses associated with or incurred during pregnancy and, because this remuneration is not for the delivery or relinquishing of a baby, it is less likely to be considered baby selling (Steinbock 1988). Nevertheless, because surrogate and gestational carrier arrangements depart so radically from traditional social norms of parenthood, the risks of objectification of women remain (Tieu 2009; Atwood 1986).

Disputes involving surrogates and gestational carriers may sometimes come to the attention of your ethics committee and you may be able to provide useful guidance to the parties involved. The questions that loom large in an ethics analysis are whether the surrogate or gestational carrier is being exploited; what emotional and psychological risks she faces; what the impact of surrogacy on the families of both the surrogate and the receiving parents is likely to be; and whether permitting surrogacy constitutes endorsement of buying and selling babies.

Termination of Pregnancy

Marlise Munoz, 33 years old, was 14 weeks pregnant when her husband found her unconscious on the bathroom floor. Rushed to the hospital, she was found to have suffered a massive pulmonary embolism and was declared brain dead. The declaration was final and uncontested by either the hospital or the patient's family. Mr. Munoz requested that, in keeping with what his wife would have wanted, the mechanical supports maintaining her organ function be removed. The hospital responded that state law required that she be kept on "life support"* until sufficient fetal development created a reasonable

---

* *Life support* in this context is misleading, since it suggests that Mrs. Munoz's life was being maintained by technological assistance, whereas, in fact, she was legally and medically dead. The importance of language, especially in discussing life and death, is addressed further in chapter 9.

chance of survival upon delivery, which would require that Mrs. Munoz's body remain connected to mechanical supports for at least an additional eight weeks. Mr. Munoz insisted that, in deference to his late wife's wishes, the hospital disconnect the mechanical supports immediately, even though he recognized that this would prevent the development and delivery of a viable baby (Ecker 2014).

What rights do the mother and the father have in this situation? What steps should be taken to preserve and support the fetus? Should the mother's body be kept on mechanical supports solely to allow a live birth?

The Legal Landscape

Roe v. Wade is the landmark 1973 U.S. Supreme Court case that legalized abortion and one of the most important legal cases in the field of bioethics. The ruling established that a right to privacy under the due process clause of the Fourteenth Amendment of the U.S. Constitution extends to a woman's decision to terminate a pregnancy. But the ruling also balanced this right against two legitimate state interests in its regulation: protecting the potentiality of life and protecting the health of the mother. The Court held that these interests become stronger over the course of the pregnancy, and it employed a trimester approach to make this more precise: for the first two trimesters, the decision is substantially that of the pregnant woman and her doctor; in the third trimester, the state may proscribe abortion in order to protect nascent human life. In a 1992 case, *Planned Parenthood of Southern Pennsylvania v. Casey*, the Court reaffirmed the essential holding of Roe but replaced the trimester framework with the "undue burden" standard intended to protect women from unreasonable barriers to abortion. A fuller discussion of the leading Supreme Court cases related to abortion appears in part IV.

Roe v. Wade prompted a national debate on the legality and morality of abortion that continues to this day, dividing the country into so-called pro-choice and pro-life camps. Whether *Roe v. Wade* will survive the many challenges to its constitutionality remains uncertain in light of the conservative bent of the current Supreme Court.

Meanwhile, however, a number of states, as well as the U.S. Congress, have attempted to make the exercise of the right to abortion more difficult for pregnant women. As of this writing, Oklahoma, Texas, and North Carolina require that pregnant women undergo and view fetal ultrasounds, along with graphic explanations of their significance (Rocha 2012). In North Dakota, providers are required to provide information about the gestational age and development of the fetus, show them a fetal sonogram, and say that, if the woman terminates the pregnancy, she will be "ending the life of a separate human being with whom [she has] an enduring relationship." In West Virginia, women considering abortion are required to undergo a vaginal sonogram, even in cases of rape. Some states require that women be told that the list of risks to the procedure includes depression, suicide, and infertility. In 2003, a Republican-controlled Congress passed the controversial Partial Birth Abortion Ban Act, which the pro-life camp defended as protecting the rights of the unborn and the pro-choice camp rejected on the ground that it constituted an unjustified infringement of the woman's right to abortion. The act was ruled constitutional by the Supreme Court in 2007 in *Gonzalez v. Carhart*. A fuller discussion of this case appears in part IV.

Ethical Issues

More has been written about abortion than any other bioethical issue and, despite the vast literature, it is a matter that continues to generate passionate and polarizing debate. Beginning in 1971 with the landmark article, "A Defense of Abortion," by the philosopher Judith Jarvis Thomson and continuing without interruption until today, articles and books have exhaustively explored every facet of the issue. The most general way of stating the ethical problem of abortion is by asking what makes it right or wrong to voluntarily terminate a pregnancy. This divides into two sub-questions: (1) what is the moral status of the fetus? and (2) why does a pregnant woman have the right to terminate her pregnancy? The issue of moral status is important because those who have moral status deserve the protection of moral norms, that is, principles and rules that state obligations and rights. Determining the moral status of a fetus or embryo does not settle the question of what may ethically be done to it, for moral status can come in degrees, but even beings with comparatively little moral status deserve moral consideration.

Several answers to when the embryo or fetus acquires moral status have been proposed. One suggestion is that moral status is acquired at conception, even before an embryo, technically speaking, has developed. Another is that the fetus acquires moral status when pathways develop to transfer pain signals from pain sensors to the brain, around 26 weeks or 6 months. Various other developmental milestones have been proposed as marking the advent of moral status, but these are not straightforward empirical determinations. Each suggested milestone has to be defended on moral grounds. To no one's surprise, consensus remains and likely will remain elusive.

Commonly accepted justifiable grounds for abortion, on which there is wide but not universal agreement, are rape, incest, and threat to the life of the mother. More controversial reasons include the mother's maturity, threat to the mother's psychological health, conflict with the mother's other life goals, and lack of support system and financial resources. Selective reduction, in which one or more embryos in a multiple pregnancy are aborted to enhance the likelihood that the remaining embryo(s) will thrive, is problematic because of the painful and difficult selection it requires. Selective abortion of fetuses with disabling conditions has generated particular controversy. According to the disability rights critique, prenatal diagnosis followed by abortion of fetuses with disabling conditions is morally problematic for a number of reasons: it expresses negative or discriminatory attitudes not only about the disabling traits themselves, but about those who have them; it also signals intolerance of diversity in both the family and society at large (Parens and Asch 2000), ultimately altering for the worse what it means to be a parent.

The pregnancy termination case of Marlise Munoz no doubt generates conflicting opinions among individuals depending on their views regarding the ethical permissibility of abortion and the reasons that might legitimate it. Questions to be asked when doing an ethics analysis include the following: What moral significance does the preservation of fetal life have? Is it one factor among others or is it the overriding consideration? What is the fetal prognosis if the pregnancy is maintained for another eight weeks and delivered? Would the mother want her body to be supported mechanically so that her baby could be delivered? What rights does the father have? Should the hospital attempt to override the father's decision to terminate mechanical supports? The goal of your ethics committee deliberations should be to draw attention to these various moral considerations and to try to come to some conclusions about their respective weights.

Maternal-Fetal Issues

Janet Jones was 32 weeks pregnant when clinical signs began to indicate that all was not right with her pregnancy. Fetal monitoring revealed a decrease in fetal heart rate that could indicate inadequate blood flow through the placenta. She was followed closely by her obstetrician during this time and, when the fetus's decelerations worsened, her doctor recommended delivering the baby by cesarean section before things got worse. Janet, however, refused. She did not believe her fetus was in serious difficulty and, in any case, she had spent her entire pregnancy preparing for natural childbirth and wanted to deliver her baby that way. Her obstetrician, Dr. DiSalvo, and other members of the obstetrical staff tried to convince her of the need for a C-section, but she continued to refuse and, as fetal condition worsened, they became increasingly worried. They felt a responsibility to do whatever they could to deliver her baby alive before it suffered irreversible brain damage. Finally, Dr. DiSalvo warned Janet that if she continued to reject his advice, he would have no choice but to get a court order to perform a C-section over her objection. The rest of the obstetrical staff, though bothered by this threat, supported his stand.

According to *Roe v. Wade*, pregnant women in the United States have a constitutional right to terminate an unwanted pregnancy at least during the first two trimesters, which is not inconsistent with their moral obligation to do whatever is necessary to ensure the success of a desired pregnancy. Women who intend to carry their pregnancy to term are advised to take care of their own health for their sake and for the sake of their future child. But, the ethics and legality of abortion are separate matters from the moral responsibilities of women who choose to carry a pregnancy to term and deliver a child. These are obligations that pregnant women have *now* to the *child-who-will-be-born*. Thus, pregnant women are advised not to drink or smoke, not to take illicit drugs, and to avoid exposing themselves to environments that present a risk to the health of the developing fetus.

Women are also expected to agree to medical interventions, such as cesarean section, that their doctors believe are in the best interests of their to-be-born child. The important distinction between a moral and a legal obligation is that, while these intuitively obvious measures can be seen as *moral* obligations, it is a further question whether they should be translated into *legal* obligations (Post 1997). As a general matter, a woman's right to privacy and self-determination cannot be legally conditioned on the well-being of her future child, but neither should she be relieved of the moral obligations to the child she intends to bring into the world.

Yet, restrictions on a pregnant woman's right of self-determination have taken several forms. Women have been ordered by courts to undergo emergency cesarean sections; incarcerated until delivery if they have taken illegal drugs while pregnant; and punished after delivery for engaging in behaviors during pregnancy that, in the opinion of doctors and courts, endangered the fetus. These actions have been justified by states modifying their child abuse and neglect laws to define a "child" as any being from conception on, thereby expanding their *parens patriae* authority to include fetal protection. These measures depart significantly from the customary legal position that personhood within the meaning of the Fourteenth Amendment occurs when a live birth takes place. Only at that point may the state, under its protective powers, intervene on behalf of the neonate, who now has in dependent and legally protectable rights (Post 1997). The seriousness of these responses to alleged maternal irresponsibility, as well as their punitive, coercive, and, often, arbitrary nature, shows how much rides on others' assessments of maternal conduct and how important it is to proceed cautiously in making them.

There are reasons to be wary of such interventions. First, obstetricians, like all physicians, are not infallible. Many of their past convictions about diet, ideal weight gain, and exercise during pregnancy are now considered obsolete. Moreover, doctors may disagree with their colleagues, as well as with their patients, about what promotes maternal and fetal health. Second, even granting the soundness of the medical advice, as noted above there is a difference between a moral obligation to do or avoid doing something and an obligation that is legally enforceable. Additional arguments, over and above the existence of the moral obligation itself, are needed to justify the use of state authority to curtail a pregnant woman's autonomy and infringe on her liberty.

Especially important in relation to maternal-fetal issues, there are pragmatic as well as moral reasons not to resort to the heavy-handed strategies described above. First, punitive and coercive responses to alleged maternal misconduct threaten to drive pregnant women away from prenatal care, with potentially worse results for the future child than if non-punitive approaches were adopted. Second, there is no bright line that demarcates punishable from non-punishable conduct. If the goal is to discourage irresponsible maternal behavior, why stop at smoking, alcohol, and illicit drugs? What about failure to have regular prenatal check-ups or take prenatal vitamins? If nothing else, this slide along the continuum of undesirable behaviors is increasingly unenforceable. From the standpoint of the future child's well-being, it is generally better to treat the pregnant woman and her fetus not as two separate entities in conflict with each other, but as a single biological and social unit with common interests (Post 1997; Rothman 1986).

There are cases, however, in which a pregnant woman's right against non-consensual bodily invasion may be limited by the overwhelming likelihood of significant and preventable harm to the future child (Chervenak and McCullough 1991). The most common example is refusal of a cesarean section where the harm to the fetus from refusing the surgery is clear-cut, imminent, and potentially devastating. What makes these cases especially difficult for caregivers is that, with surgery, this almost-baby can be saved. The case of Janet Jones may be an example of this type. In some cases, court orders have been obtained to authorize the contested intervention and, as an example, *In re A.C.* is discussed in part IV.

Given the well-settled right of capable patients to refuse unwanted treatment, discussed in chapter 3, this radical departure from legal and ethical precedent has ominous implications. It may have the counterproductive effect of discouraging pregnant women from seeking prenatal care out of fear that they might be forced to undergo unwanted surgery for the sake of their fetus. As noted in chapter 2, refusal of recommended treatment does not, by itself, confirm decisional incapacity. Yet, in these situations, surgery over the patient's objection is sometimes justified by the presumption of incapacity, the logic being that no rational woman would deliberately put her future child at risk. As "A Defense of Abortion" so brilliantly illustrates, it is hard to imagine analogous actions of non-consensual bodily invasion, especially for the sake of another, occurring in any other patient population or clinical setting. An obstetric case that raises the possibility of court-ordered cesarean section might come to the attention of your ethics committee and your careful analysis of the relevant rights, interests, and obligations will be crucial to its resolution.

Prenatal/Newborn Genetic Testing and Genomic Newborn Screening

The increasing ability to detect actual or potential medical problems enables parents and care professionals to intervene in ways that have profound health benefits. To appreciate the ethical implications, however, it is necessary to distinguish two types of assessment: *testing* and *screening*. Prenatal genetic testing is offered to individual prospective parents, generally on the basis of family history, to identify a specific genetic variant or mutation in their offspring. Genetic testing of newborns may be offered on the same familial basis or because suggestive symptoms have already appeared. Currently, this testing is undertaken only for identifiable *early-onset* conditions, the early diagnosis of which can lead to interventions that have therapeutic value. Newborn testing for *late-onset* conditions or for medical conditions for which there is no cure or other type of medical benefit is ethically problematic and may be objectionable because the child may be able to live a normal life for many years before the onset of disease. Also problematic is prenatal genetic testing for the purpose of identifying fetuses with disabilities or minor medical conditions, although this might be defended as an exercise of reproductive freedom. In a more speculative and ethically contentious vein because of its eugenic overtones, there is the use of prenatal genetic testing to create so-called designer babies by identifying genes that are not associated with medical conditions but with desirable traits, such as superior intelligence and memory.

The aim of newborn screening is different. As currently practiced, screening is intended to identify newborns from a particular population who could be helped if their heightened risk of a specific disease were recognized. Parents could then be offered follow-up genetic testing or some type of therapeutic recommendation.

In contrast, whole genome screening (WGS) of newborns, which examines the entire genome, raises ethical and social issues that targeted newborn genetic testing does not. It is now technically feasible to analyze a newborn's entire genome to reveal her genetic variations. One concern, however, is cost: WGS is currently prohibitively expensive for general use. More important, the clinical utility of newborn genomic screening is questionable, given the current state of scientific knowledge. Unlike the relationship between genetic variation and monogenetic diseases, the relationship between genetic variation and polygenetic disorders, which comprise the majority of human genetic diseases, is still not well understood, making it difficult to interpret and, therefore, make diagnostic, prognostic, or therapeutic use of this information.

As to ethical issues, although there is some overlap between those raised by newborn testing and screening programs and genomic newborn screening, the latter raises distinct problems because of the wide net it casts. WGS can provide information on many genetic variants whose significance is not understood and that may not be linked to disease or significant impairment. Or they may be linked to late-onset conditions, such as Alzheimer's disease, or to conditions, such as Huntington disease, for which there is no cure or ameliorative intervention. The danger is that parents anticipating medical problems in their basically healthy children might subject them to interventions that are unnecessary, burdensome, and even damaging. In addition, the psychological impact of this information on young people could be extremely traumatic and disruptive to their lives.

To be sure, genetic testing of newborns raises some of the same ethical issues. There are questions about what the consent process should be and who should be able to provide consent, the impact of the test results on children and families, safeguarding the privacy of the genetic information revealed, and preventing discriminatory repercussions in employment or insurability. But genomic newborn screening magnifies the double-edged implications, potentially providing vast amounts of useful or unnecessary information to health professionals and families, as well as the risk of creating a "medicalized society" (Almond 2006). On the positive side, newborn genomic screening could offer important diagnostic, prognostic, and therapeutic tools, potentially leading to disease prevention, early intervention, and the development of more effective medicines that are tailor-made to a child's specific medical condition.

An additional concern is the effect that the availability of WGS may have on notions of good parenting and parents' procreative decision making. Given the widely advertised possible benefits genomic information might provide, parents may be susceptible to the notion that they have a duty as good and responsible parents to take advantage of this type of screening (Donley, Hull, and Berkman 2012). The corresponding duty is that of the genetics professionals who, in addition to providing accurate information, also provide explanation, interpretation, and counseling. From an ethics perspective, analysis should focus on the potential benefit-burden-harm ratio; the interests and vulnerabilities of the child and the parents; and the obligations of health care professionals to provide information, support, and guidance in making decisions with profound and lasting implications.

References

Almond B. 2006. Genetic profiling of newborns: Ethical and social issues. *Nature Reviews/Genetics* 7:67–71.

Atwood M. 1986. *The Handmaid's Tale*. Boston: Houghton Mifflin Company.

Baily MA, Murray T. 2008. Ethics, evidence, and cost in newborn screening. *Hastings Center Report* 30(3):23–31.

Brezina PR, Zhao Y. 2012. The ethical, legal, and social issues impacted by modern assisted reproductive technologies. *Obstetrics and Gynecology International*: 1–7.

Brock D. 1995. The non-identity problem and genetic harms: The case of wrongful handicaps. *Bioethics* 9(2):269–76.

Centers for Disease Control. 2012. Assisted Reproductive Technology National Summary Report, www.cdc. gov/art/ART2012/NationalSummary_index.htm (accessed August 2014).

Chervenak FA, McCullough LB. 1991. Justified limits on refusing intervention. *Hastings Center Report* 21(2):12–18.

Congregation for the Doctrine of the Faith. 1987. *Instruction on Respect for Human Life in Its Origin and on the Dignity of Procreation: Replies to Certain Questions of the Day*. Rome: Vatican.

Daniels CR, Heidt-Forsythe E. 2012. Gendered eugenics and the problematic of free market reproductive technologies: Sperm and egg donation in the United States. *Signs* 37(3):719–47.

Dickens BM, Cook RJ. 2008. Multiple pregnancy: Legal and ethical issues. *International Journal of Gynecology and Obstetrics* 103:270–74.

Donley G, Hull SC, Berkman BE. 2012. Prenatal whole genome sequencing: Just because we can, should we? *Hastings Center Report* 42(4):28–40.

Ecker JL. 2014. Death in pregnancy: An American tragedy. *New En gland Journal of Medicine* 370(10):889–91.

Epker JL, de Groot YJ, Kompanje EJO. 2012. Ethical and practical considerations concerning perimortem sperm procurement in a severe neurologically damaged patient and the apparent discrepancy in validation of proxy consent in various postmortem procedures. *Intensive Care Medicine* 38:1069–73.

Ethical and Religious Directives for Catholic Health Care Ser vices. 2009. 5th ed. United States Conference of Catholic Bishops.

Ethics Committee of the American Society of Reproductive Medicine. 1999. *Fertility and Society* 72(4).

Goldenberg AJ, Sharp RR. 2012. The ethical hazards and programmatic challenges of genomic newborn screening. *Journal of the American Medical Association* 307(5): 461–62.

Malek J, Daar J. 2012. The case for a parental duty to use preimplantation genetic diagnosis for medical benefit. *American Journal of Bioethics* 12(4):3–11.

Murtagh GM. 2007. Ethical reflection on the harm in reproductive decision-making. *Journal of Medical Ethics* 33(12):717–20.

National Center for Health Statistics. 2011. *Assisted Reproductive Technology (ART): Section 4: ART Cycles Using Donor Eggs,* www.cdc.gov/art/ART2011/section4.htm (accessed February 2015).

Orr RD, Siegler M. 2002. Is posthumous semen retrieval ethically permissible? *Journal of Medical Ethics* 28:299–303.

Parens E, Asch A, eds. 2000. *Prenatal Testing and Disability Rights.* Washington, DC: Georgetown University Press.

Parfit D. 1984. *Reasons and Persons.* Oxford: Oxford University Press.

Post LF. 1997. Bioethical considerations of maternal-fetal issues. *Fordham Urban Law Journal* 24(4):757–75.

Rhoden N. 1987. Cesareans and Samaritans. *Journal of Law, Medicine & Health Care* 15(3):118–25.

Richards SE. 2014. The next frontier in fertility treatment. *The New York Times,* January 12, p. A21.

Robertson J. 2003. Extending preimplantation genetic diagnosis: Medical and non-medical uses. *Journal of Medical Ethics* 29:213–16.

Lewin, T. 2014. Surrogates and couples face a maze of state laws, state by state. *The New York Times,* September 14, p. A1.

Rocha J. 2012. Autonomous abortions: The inhibiting of women's autonomy through legal ultrasound requirements. *Kennedy Institute of Ethics Journal* 22(1):35–58.

Rothman BK. 1986. When a pregnant woman endangers her fetus. *Hastings Center Report* 16(1):24–25.

Saul S. 2009. Birth of octuplets puts focus on fertility clinics. *The New York Times,* February 12.

Steinbock B. 1988. Surrogate motherhood as prenatal adoption. *Journal of Law, Medicine & Health Care* 16(1):44–50.

Suter S. 2011. Bad mothers or struggling mothers? *Rutgers Law Journal* 42:695ff.; 35(3):171–75.

Thomson JJ. 1971. A defense of abortion. *Philosophy and Public Affairs* 1(1):47–66.

Tieu MM. 2009. Altruistic surrogacy: The necessary objectification of surrogate mothers. *Journal of Medical Ethics* 35(3):171–75.

DISCUSSION QUESTIONS

1. How could prenatal genetic screening be used for eugenical purposes?

2. If you had the ability to make decisions about your offspring's traits, would you do so? (Some traits might include hair color, hearing ability, skill in sports or activities, likelihood for cancer, etc.)

3. How does an individual's economic status affect their ability to have a child?

4. What is the relationship between ethics and politics regarding reproductive rights?

5. If IVF/surrogacy is legal but unaffordable, should we still say such services are accessible?

6. Why do some private matters become political issues?

7. What rights ought a fetus have?

8. How does the relationship between money and bodies affect the way we understand surrogacy?

Recommended Cases

Marlise Muñoz—Although she was declared dead, Marlise Muñoz's body was kept on organ support, in accordance with state law, so that the 14-week-old fetus she was carrying could have the possibility of being viable.

Giselle Marrero—After expressing her husband's desire to have children, Giselle Marrero was given a court order allowing the extraction and preservation of her dead husband's sperm. This was done through a process of postmortem sperm retrieval, which has occurred since 1978.

China's One-Child Policy—Beginning in 1979 and concluding in 2015, when it was changed to a two-child limit, China's one-child policy limited the number of children allowed per family to one (with some exceptions). To enforce the policy, the government could require contraception, sterilization, and abortion, in addition to fining violators heavily.

UNIT V

Experiences

....

Unit V Introduction

In bioethics, we shall come across individuals in a variety of populations and situations. Their circumstances may be dictated by medical structures, such as the doctor–patient relationship, or from having different bodies, histories, or backgrounds. A bioethical examination is incomplete if it does not address the shortcomings evident in medicine and medical research. The history of medical segregation is plentiful, but the impact of such history requires investigation. Uncovering the consequences of an unethical past pushes the field forward.

This unit attempts to answer one question: How do we understand the experience of another person? The salience of this topic comes from a corollary: the experiences of another person matter in decision-making. To focus on one example, a patient with an illness is more than their chart; there is an existential impact of illness. If a doctor refuses to recognize these impacts, the patient can be harmed. A phenomenological understanding of such illness would be required to uphold the Hippocratic duty to treat a person, not an illness or a chart. By prioritizing experience, the patient, the doctor, and others involved in caretaking take priority, rather than a philosophical or medical abstraction. It ensures that we are engaging with another person, rather than reducing them to their position in the relationship.

This unit also focuses on the differences between people and the medical lacks caused by such inferences. For instance, the majority of medical research begins with a male body, and this results in women's particular needs being omitted. One deadly instance is the different symptoms of a heart attack. Until recently, it was thought everyone had the same symptoms. We now know women and men have different symptoms, but the historical framework of heart attacks is hard to escape. Currently, women are seven times more likely to be sent home from the Emergency Department while having a heart attack.

"The Law of the Body" focuses specifically on experiences differentiated by bodies. Holloway focuses on bodies that have been underrepresented in the histories of medicine and bioethics. By examining the role medicine had in corrupting these bodies, we can explore the possibility of a more ethical treatment of all bodies and of all people, especially those that have been historically mistreated. There is an emphasis on the right of privacy and how privacy has been taken from these individuals' hope for autonomy.

The focus of "Attending to Difference: Phenomenology and Bioethics" is the relationship between bioethics and the philosophical field of phenomenology. Phenomenology is the philosophy of the structures of experience, or, rather, of consciousness and the objects of consciousness. The focus of phenomenology is that of the individual experiencing the world. This paper focuses on the critical relationship between bioethics and phenomenology as a method of connection and insight and for progress.

In "Toxic Life in the Anthropocene," we encounter the environmental impacts on our experience. Through a diverse group of case studies, we will examine the causal links between our world and our medicalized existence. Using a global survey, we see the bifurcation of nature and nurture undermined to view the ways our world permeates our health and ourselves.

The Law of the Body

Karla FC Holloway

....

Privacy's Privilege

Terri's Law • After Terri Schiavo had been in a persistent vegetative state for nearly fifteen years, her husband, Michael, attempted to end her prolonged dying. A respirator and a feeding tube had kept her breathing and provided her body with sustenance following the respiratory and cardiac arrest that she experienced in her Florida home in 1990. Michael sought to end the artificial life support, a decision he felt his wife would have made for herself had she been competent to do so. After years of fruitless efforts to alter his wife's state, he finally petitioned for removal of the medical devices and, as her legal guardian by marriage, he attached a do not resuscitate order to her medical records. His decision was opposed by her parents, and the court battles that ensued to contest her parents' attempts to halt the removal of the tubes brought publicity to the case. In an extraordinary move for such a private family matter, her parents' efforts were eventually supported by the Florida legislature, which passed "Terri's Law"—legislation that effectively gave Jeb Bush, then the state's governor, authority to intervene in the case.

An appeals court ruled that Terri's Law was unconstitutional. But before that ruling, there was an unprecedented public intervention in what one might have reasonably considered a private family matter. Following the verdict by Florida's Second District Court of Appeals to adhere to her husband's decision to remove the feeding tube—a decision fought by her parents—the U.S. Senate passed legislation that allowed it to enter the family's quarrel, and a bill called the "Palm Sunday Compromise" was introduced in the House of Representatives, requiring that the Florida case be moved from state to federal jurisdiction. Outside of the nursing home where Terri lay dying, Randall Terry, a right-to-life activist, orchestrated noisy public demonstrations to protest Michael Schiavo's efforts.

The legions of protestors outside of the nursing home ensured that Terri's dying would be a public spectacle. Michael's decision to end her life ultimately prevailed, and Terri Schiavo died on March 21, 2005. On her tombstone are two death dates: 1990 and 2005. Too many people know the intimate details of what might have been a private family tragedy.

Henrietta's Haunting • On the day in October 1951 when Henrietta Lacks died of ovarian cancer in Baltimore, Maryland, her husband and children completed the necessary paperwork to release her body from the hospital morgue and sorrowfully began to make the funeral arrangements that would return the body of their wife and mother to her home for burial. The grief-stricken family was fully engaged in the business of her burial. They had no idea that on that very evening, Dr. George Gey, a tissue culture researcher at Johns Hopkins University, where she had been a patient and had died, had convened a press conference. Clutching a test tube that contained cells from her tumor, cells that his laboratory had collected months earlier, he announced that a new era of medical research had begun—one that, someday, could produce a cure for cancer. The family knew nothing of Gey's television appearance, nor had they any idea that cells had been harvested from Henrietta's tumor.[1]

Some twenty-five years later, Barbara Lacks, the wife of one of Henrietta's sons, was invited to a friend's home. The friend had also invited her sister and brother-in-law to dinner. The brother-in-law was a scientist from Washington, and after introductions and some conversation, he reportedly turned to Barbara and said: "Your name sounds so familiar ... I think I know what it is ... I've been working with some cells in my lab; they're from a woman called Henrietta Lacks. Are you related?" "That's my mother-in-law," Barbara whispered, shaking her head, "She's been dead almost twenty-five years, what do you mean you're working with her cells?"[2]

Even before Henrietta Lacks succumbed to the ovarian cancer that had spread throughout her body, the cells from the tissue culture taken from her tumor were already exhibiting an extraordinary vitality. In the first quarter-century after her death, her cell line had become legendary in research labs around the world. The sociobiologist Hannah Landecker perceptively explains the vitality of the cell line (known as HeLa) developed from Henrietta Lacks's tumor. Landecker writes: "HeLa continues to be used, explained, exploited and narrated in the scientific and popular press, as well as through film and television, making it one of the most *storied* biological entities of the twentieth century" (emphasis added).[3] Landecker's acute understanding of the unusual circulation of this event is reflected in her vocabulary: she notes "the personification of the cell line in the image of the woman from whose body it was extracted." That night over a dinner table, two worlds collided. One was private and involved a family who still mourned the death of their mother. The other was public and concerned a scientific community that had been using her tissue cultures to grow cell lines for a wide variety of medical research. Until that moment and the later publicity that came to be attached to Henrietta Lacks, the family had borne their loss privately, and medical research had profited from the vitality of a cell line that became immortal (another personification) and the product of Microbiological Associates, Inc., the biological supply company that made the cells available for commercial use.

---

1 Skloot, "Henrietta's Dance."

2 Ibid. Skloot continued her interest in this subject and published *The Immortal Life of Henrietta Lacks* in 2010.

3 Landecker, *Culturing Life*, 160.

The robust utility of the tissue sample taken from Henrietta Lacks was indeed extraordinary. Like no other human cellular material before it, the cervical cancer cells that came from Henrietta Lacks had a remarkable vitality. Before the recovery and use of her tissue samples, there were no immortal human cancer cell lines—that is, there were no human cells that did not eventually experience cell death, or apoptosis. Mrs. Lacks's cells did not die. The science reporter Michael Gold describes what happened to the cells from her tumor:

> At some point, the cells in each little island of tumor began to quiver and dance … and multiply. Where there had been one, there were now two. Where there had been two, now four, now sixteen, now thirty-two. In a few days, the signs of growth were visible: around each cube [of extracted tissue] a translucent band of new cells was taking shape … Every other human cell line in the Gey lab had faltered. Yet as the months passed, the HeLa cells showed no such vulnerability. They just kept growing, doubling their number every twenty-four hours … the tumor cells were growing ten to twenty times faster than the normal cervical tissue taken from Henrietta Lacks.[4]

George Gey, the Hopkins scientist who was head of the tissue culture lab, had given instructions that all cultures taken from cancer patients should be presented to his lab for analysis. An article in the *Johns Hopkins Magazine* described the sample taken from Henrietta Lacks: "They multiplied like nothing anyone had seen. They latched to the sides of test tubes, consumed the medium around them, and within days, the thin film of cells grew thicker and thicker."[5] The discovery was extraordinary and changed the field of medical research. But the same aggressive growth that appeared in the laboratory attacked Henrietta Lacks. Her cancer spread to almost every organ of her body, and nine months after she was diagnosed, she died. Until that coincidental dinner table conversation some twenty-five years later, her family had no knowledge that Henrietta's biological material had entered the public domain and had become the stuff of medical legend.[6] However, many thousands of medical researchers knew of the HeLa cell line's extraordinary ability to multiply. And some, like the dinner table guest, knew Henrietta Lacks's name.

George Gey's interest in keeping that name private had been overridden by Roland Berg, director of scientific information for the National Foundation for Infantile Paralysis, who had professed his

4 Gold, *A Conspiracy of Cells*, 18.

5 Skloot, "Henrietta's Dance."

6 Van Smith, "Wonder Woman: The Life, Death, and Life after Death of Henrietta Lacks, Unwitting Heroine of Modern Medical Science," *Baltimore City Paper*, April 17, 2002 (http://www.citypaper.com).

expertise in working with the public with this "type of material" and had told Gey that "you do not engage the attention of the reader unless your story has basic human interest elements."[7]

The Washington scientist invited to dinner with Henrietta Lacks's daughter-in-law knew the name of the source of the cell line that made his research possible because there had been a decided effort and specific intent to attach narrative elements to this unusual medical story, despite the fact that the Lacks family was never told that her cells were available to medical researchers. Landecker perceptively weaves the critically constructed narrative into the scientific utility of the cells:

> The widespread presence of the cells in laboratories was equated with the ongoing, if distributed, presence of the woman's life essence. In this period, the personification of the HeLa cells in the figure of Henrietta Lacks was a beneficent one, a story of unfortunate death turned to the benefit of mankind in conquering polio. The cells were understood to be a piece of Henrietta Lacks that went on growing and dividing, encased in a test tube instead of a body ... Their concomitant personification was in the form of an angelic figure, an immortalized young Baltimore housewife, thrust into a kind of eternal life of which such a woman would never dream.[8]

It was also an "eternal life" that had not been officially revealed to her heirs and descendants. The rest of the text of this narrative is invested with the consequential remnants of her racialized identity. Her "angelic" persona fell away when the cell line was somehow contaminated, likely in one of the many laboratories that circulated the material. Following this new reference for the HeLa line—that it was not a pure sample but a contaminated one—that same personification that attached to her altruistic (albeit unknowing) gift of her body's most intimate material, Lacks's own identity—understood fully as a black woman—was rendered vulnerable to the social stereotypes habitually attached to race and sex. The vocabulary that described the HeLa cell line as "vigorous," "aggressive," and "difficult to control" was not detached from the body that produced them. Stereotypes of black women have historically used that kind of imagery.[9] The disciplined image of Mrs. Lacks that appears with the discussion of HeLa cells has incorporated the very stereotypes—of authoritative power and aggressive dominion—that black women of her era struggled against.

---

7 Landecker, *Culturing Life*, 163.

8 Ibid., 164.

9 See, for example, the literary scholar Trudier Harris's essay "This Disease Called Strength," which explores the evolutionary stereotype of the overly strong, aggressive, indomitable black woman.

Intimate Interventions and the Public Domain

Among the most intimate, most extraordinarily private moments of our lives are those that involve contraception and birth, medical interventions that through experimental or standard care seek to alter the quality of our lives, the event and occasion of our dying, and when we learn more about our biological lives—the remarkably personalized data that is available through genomics and discoveries of the unique information about ourselves that is attached to our DNA. Each of these areas—broadly engaged in this book's four chapters, which focus on reproduction, genomics, clinical trials, and death and dying—has the potential to have an inordinate effect on the ways in which we manage our daily lives. These are experiences that seem universal. But they are decidedly particular as well. Because of the unique cultural and social histories attached to specific populations in the United States, each of these areas has an impact on women and on minorities that inevitably questions our presumptions regarding individual autonomy. At their root, matters of identity are attached to the cultural ideologies regarding a person's fundamental right to privacy.

Private Personhood • In the United States, privacy has a value that is intimately attached to the history of our laws, as well as to the social evolution of our attitudes regarding difference. Gender and race have particular (arguably, peculiar) legal histories in the United States that make the idea of privacy read differently for different bodies. Perhaps as a consequence, or merely in concert with the evolution of differential treatment in the law, our social judgments and cultural perspectives leak into medical practice as well as into advances in science. Of course we are eager to claim and to want science and medical research to be neutral, to concern themselves with facts rather than the politics of identity. But this book argues that social judgments and social systems are critical dimensions of science and medicine. In fact, as medicine reaches toward genomic science for information that might be applicable to individualized therapies, neutrality becomes an even more complicated terrain. Medicine mixes knowledge gained from science and research and knowledge gained from social interaction with a patient.[10] As human as the right to privacy might seem, it has a public history that absolutely renders it a socially selective privilege.

One way to consider the idea of privacy as it is related to the issues of reproduction, genomics, clinical trials, and death and dying is to understand how each of these issues involves the ethics of identity. There is reasonable debate regarding the freedom we have to select our identities, an argument that engages the concept of liberal personhood that John Stuart Mill would champion. But identity is not an insular matter. It is fully shaped by the social narratives that adhere to certain bodies, without concern for or interest in the individual desire for sameness or difference that an autonomous selfhood would assert. Within this frame, sociality is determined by the strength of the identitarian narrative. The most focused consideration of this book is directed toward what it is about

---

10 One of the most obvious problems for genomic science that might predict certain disease outcomes based on race is an African American patient whose race is socially constructed, and who would reflect a genomic history that indicates the complicated social histories of racial interactions in the United States.

race and gender that goes beyond liberal personhood and that controls how identities, rather than persons, interact within the public sphere.[11]

The argument of *Private Bodies, Public Texts* is that narratives—socializing stories—that are attached to all women and to blacks of both genders have an inordinate control over the potential for private personhood. The public controls of race and gender are so robust that private individuation is rarely an opportunity for those whose identities fall within these two social constructs. My consideration of privacy as the ideal but vexed realm of individual personhood explores the ways in which individuality is either subsumed into this nation's social history or extracted from it. The consequence is the compromise of privacy's privilege. We find clear examples of this loss in the ways in which law adjudicates, and medicine practices, disciplinary and regulatory control. The most intimate aspects of our being and the stories that reveal them give shape and contour to a lens that ultimately reveals the loss that women and blacks in the United States experience, as they are woven into collective social categories that separate them from the privileged privacy inherent in the presumptively white and male ideal of liberal personhood.

This book exposes the capaciousness of this loss within legal and medical cases, from legal theory and bioethics. When the legal scholar Radhika Rao—in an essay on law, medicine, and ethics—poses a question regarding the location of law that might touch property (a legal realm that once included the bodies of enslaved persons), contract (who might enter into such a legal relationship, and who was excluded), and privacy claims, she affirms the narrowed personhood that situates her argument, asking: "Which body of law should become the law of the body?" Rao recalls the case of the Californian John Moore, whose claim to his spleen was denied in a well-known case that held that persons may not claim remnants of their body once it is separated from them. Concerning *Moore v. Regents of the University of California*, Rao writes that "Moore's description of the wrong done to him is quite revealing: 'My doctors are claiming that my humanity, my genetic essence, is their invention and their property.'"[12]

Rao needed Moore's story, as well as his words, to pose her legal questions. Nevertheless, the legal issues were settled without regard to Moore's claim to the integrity of his body as fully human. Other than in depositions or lower court testimonies, a plaintiff's words are not ordinarily found within the decision of an appeals court. At the appellate level, the legal question reigns, and the voices that represent that question are less important than its reach. But Rao uses the voices to great effect in her essay—because the case law insufficiently mined the layered social and cultural narrative that would expand the story. Its complexity was not sufficiently apparent through the plain facts, the legal questions, or the four corners of the final decision.

---

11 This book's focus is on the United States, so "race" for blacks living in the United States will be captured with that imperfect nomenclature. It does not mean to dismiss the wide differences of geographies dispersed within the African diaspora—the Caribbean, Europe, and Canada, for example—that are held within the term *race*.

12 Rao, "Genes and Spleens," 377.

Moore has reached near legendary status in law and bioethics and is now arguably an ur-text of medical ethics and privacy law. But one of the reasons for its utility is the story's subtext. At least a part of its seminal distinction is that Moore's body was one a reasonable person—the normative standard of the law—might have presumed did indeed belong to him. His gender and his race (he is a white male) placed him among society's most privileged and outside of the coded social identities that have an embedded vulnerability. However, the issue that made the privilege of a private identity *not* the case in Moore's situation—the state's interest in the advance of medical knowledge—is unquestionably worthy of the sustained attention that this case has received in biomedical ethics and property law. Equally significant is the way in which Moore's unexpected occupancy of a privacy and property claim is a telling acknowledgment that there are some bodies who seem to have this entitlement as a normative protection that extends from their citizenship, and there are other bodies who are not only denied this presumption of privacy, but who cannot reasonably anticipate its protection. These other bodies have names and case histories that, even if they are familiar, have not been read into the literature of medical ethics or into the histories of case law within the context of their embodied politics.

My argument in this book is that these other bodies—women and blacks—are bodies that begin with a compromised relationship to privacy. The essential difference between them and John Moore is that their experiences attach to race and gender in ways that Moore's did not. We do not need to know Moore's race or gender in order to read the case, to understand his claim, or to determine its outcome. In law and bioethics reviews, Moore is an individual who was denied a property claim to his body. But those other bodies that make their appearances in law and medicine as distinct and memorable cases, like Terri Schiavo's, often do so through the veil of gender or race. The social categories they occupy do not disappear in objective analyses. In fact, their very subjectivity disrupts their claims to an individual, liberal personhood.

The experiences of women and black Americans are particularly vulnerable to public unveiling. We cannot appreciate the substance of these intimate violations without the accompanying texts of the social identities that govern them. For example, Terri Schiavo was rigorously gendered throughout the frenzied public attention to her plight. She was persistently defined through her familial relationships (as a wife and a daughter), as if those categories by themselves shaped the lens through which the public might more clearly see the dimensions of her tragedy. Privacy is presumed to be a fundamental right of personhood. Nevertheless, particularly for those who made late appearances into constitutional protection, that legal notice has become an assurance that some corporeal feature is always and already vulnerable to public attention as an identity that stands in the place of private personhood. It is as if the Fourteenth Amendment's assurance of equal protection has come to mean that women and blacks would be noticeable first through the legal identity they constitutionally occupy, and only later, and only perhaps, as private persons. What is an appropriate perspective for this vulnerable personhood? One is to acknowledge the thick narratives that reside within the public and the private.

Literary fictions reveal the deeply textured terrains in which stories (or cases) reside. The complex landscapes from which fictional characters emerge might stand in for the texts of fuller narratives that do not often (or appropriately) make it into case law or case histories. The stories that lie between fiction and fact make public narratives resilient. Although law and medicine are rigorously practical fields not much given to the imaginary, literature's stories can encourage our substantive notice and consideration of the ethics embedded within both disciplines. The extended contexts and considered complexities that help us unveil the ways in which privacy and identity matter regarding our perspectives of some bodies benefit from the texture of fiction. Unlike law and medicine, fiction is free to indulge its complications rather than constrained to rigorous enforcement of boundaries that encourage expeditious resolution.

There is good evidence that some of the most compelling fictions use story lines linked to medicine and law. In popular culture, this is why courtroom and hospital dramas so liberally populate some of television's better-known shows. Students in my undergraduate class called Bioethics and Narrative often notice that an issue we are discussing in class appeared in a television show or a film they recently saw.

In the 1960s, television's early heyday, the stories of doctors Kildare and Marcus Welby captivated viewers' attention. In the latter decades of the twentieth century and well into the first decade of the new century, some of television's most popular dramas have been about medicine and law. Audiences are captivated by the issues, characters, and story lines that often take their texts from real-life events. But literary fictions make possible a more complicated life for the legal subject or the medical body than that normally presented in depositions or medical case histories. This more nuanced consideration of complexity is a benefit, especially in fields that try to resolve an issue with expediency. This book's engagement with literary fiction as supplementary material for the ethics-focused analyses of legal and medical narratives is in part specifically because cases in law and medicine are contextually slender. The professional frames of those fields appropriately encourage stripping the story to the essence that results in a lament like John Moore's, that medicine found his humanity in his body's molecular essence rather than in his personhood. In my judgment, our reasoned consideration of the issues that these narratives present stands in need of the nuance that returns the bodies under discussions to full personhood rather than rendering them as essentialized subjects. Fiction is an appropriate and richly textured source for the fully human.

A Patient or a Body? • A physician's comment to me following a talk I gave to the medical humanities group at the Baylor College of Medicine called "The Body Politic" has lingered long past the moment of that lecture. He explained that his aim in teaching was to encourage his students to see patients as people, rather than the disarticulated "bodies" that constituted the topic of my lecture. Bodies are what we have in pathology labs, he explained, but then he went further: "I want my students to understand their patients as human beings—not just human bodies."

I have thought about his comment a good deal and wondered whether it was my inclination, as a cultural studies scholar, to use the language of "the body" in a way that actually belied the point I was making—that the social and cultural have an unwarranted impact on the bodies that are the focus of our medical judgments and the ethics of care.

Both the process and subject of this book are in part a response to the potential of disciplinary divides to separate the subjects of medicine, law, and ethics that would otherwise, it seems to me, find their coherence in a consideration of the body. Instead of distinguishing the legal subject from the patient, and the physician and lawyer from both of these, I want to integrate these subjects into multiple narrative landscapes that examine a dialogic constitution of identity.[13] The result of my effort is this interdisciplinary discussion and focus for this book, with law, medicine, and ethics in conversation with each other and bounded by the thick frames of cultural and literary studies.

Although the shared goal of medical ethics and legal analysis is to find an executable solution to an immediate claim or event, they do so from locations which are both complex and contradictory. The disciplinary approach to legal or medical queries is to pare down the subject to an actionable space, to reduce the scene or event to the legal question or the medical necessity. Whether we appoint an ethics committee to resolve a medical issue or a jury to settle a legal question, we contain these matters within disciplinary constraints that ensure the production of familiar and usable results. Certainly the efficiencies urged in medicine and law are appropriate and necessary; but efficient problem solving is too often achieved by a blind stripping away of material that is textually rich even though it is determined to be "excess" to the cases. I want to urge that, at the very least, we know the excess that we peel away. The messy and unexpected dimensions of fiction might encourage our deliberate engagement with the intricate dimensions of story.

Although the field of narrative bioethics emerged precisely to engage these matters, its focus on patients, physicians, and medical professionals creates a selective universe of actors that influences the judgments and perceptions that patients and medical professionals bring into the arena of medical care. Even when a patient's narrative appropriately occupies the privileged interrogative space, becoming a close reader of—or a more attentive listener to—a patient's story effectively situates that story as a coherent document from which a reading might yield a similarly coherent ethical judgment.

The narrative interest in this book is less about discovering a coherent story which might yield a usable text for a medical professional to read with close attention. As reasonable and well-intentioned as it is to employ "teachers of literature, novelists, storytellers, and patients who have written about their illnesses ... to attain rich and accurate interpretations of these stories, and to grasp the plights of patients in all their complexity,"[14] another perspective might suggest that there is an inherent tension between the goal of accuracy and the landscape of complexity. Fiction is made up of complexity. It gains its substance from engaging multiple, even contradictory, meanings. Accuracy—the absolutely

13 See Appiah, *The Ethics of Identity*, for a compelling analysis of identity, autonomy, and ethics.
14 Charon, *Narrative Medicine*, 3.

appropriate goal of law and medicine—is challenged in the fictive environment, and it is the judgment of this book that this challenge is reasonable precisely because of the facts that compose a fictive imagination. The interest of this book is to shift the direction of narrative bioethics away from the disciplinary boundaries that receive a patient's story. Imagine, for example, the common scenario at the doctor's office: "Are you in any pain today? What is the level—from 1 to 10—of your pain?" The questions do not allow for the complex of heartache, the pain of one's spirit. What number might one assign to that?

This book suggests that fiction's sometimes incoherent, messy solutions (if there are solutions at all) and its general tolerance for complexity constitute a narrative that actually fits the rest of the text that a medical professional, no matter how patient a listener or how close a reader, will not be likely to hear from a patient—especially as the questions asked are so strictly regimented. So instead of assigning relative weights to stories told by doctors, nurses, and patients, this book's approach is to look outside of those boundaries for a form of narrative that creates its full and complex stories from the very dilemmas that are capsules in patient care.[15] Additionally, the book interrogates the histories in medicine and law, the cultural texts, that contribute to the stories that narrative bioethics would assign to the patient.

On Privacy, Identity, and Ethics • How does a legal notion of privacy find its way into a cultural narrative? How can privacy articulate a relationship to identity? In bioethics, the traditional engagement with ethics descends from establishing relationships to overarching moral principles (justice, autonomy, beneficence, and nonmaleficence) that moderate and negotiate the solutions that ethical crises demand. These principles encourage our review of ethical quandaries by linking them to the conflicts that might be occasioned in particular circumstances. For example, when the wishes of a patient express the patient's desire for more aggressive care but conflict with a physician's judgment of the futility of further medical intervention, which ethical principles are proffered as the forum for resolution of the dispute? Who is the legal parent—the woman who gestates or the woman who produces the egg? Does race matter? If there is a difference between the race of the zygote and that of the gestational woman, does race matter? Should a woman use selective reduction to choose an embryo that might develop into a tissue match for a child with a terminal illness, who might be saved by the birth of a sibling who could be a donor? Each of these questions takes a private event of the body—illness, pregnancy, or death—and renders it public. The privacy ordinarily anticipated and earned by the intimacy of the event vanishes as certain questions emerge. When there is a public rehearsal of issues, what is the consequence for the bioethical principle of autonomy and the legal right to privacy? What association exists between autonomy and privacy?

These questions involve medical ethics and legal precedents. Medicine and law have designed protocols and doctrinal analyses to regulate the ways in which a solution might be shaped for the problem they represent. Many of these solutions seem finally unsatisfactory to me for two reasons.

---

15 For a view of narrative bioethics, see ibid.; Charon and Montello, *Stories that Matter*; and H. Nelson, *Stories and their Limits.*

First, there are cultural questions that can attach to these matters—those that negotiate between privacy and identity—and that are entangled in the solutions. Second, the solutions look outward, toward an objectified subject, rather than inward, at the subjective construction of the discipline. When we utilize the focused perspectives of law or medicine to resolve the dilemmas they represent, matters of culture or identity are either so overregulated that they are stripped of sense, or they disappear into a specific protocol of analytical inquiry.[16] But the textured narratives that are embedded within seem to give both agency and authority to the bodies (disciplinary and corporeal) that center each inquiry. What social judgments about personhood are embedded in the recitation of these stories? How does disciplinary genesis matter when we consider liberal personhood?

Identity matters when private personhood is made public. In a public sphere, where gender or race reign through the force of social construct, historical pattern, and even constitutional authority, social narratives are shaped within the nation's body politic. It is my argument that there are some bodies that will and can ordinarily disappear into the normative, that are not vulnerable to the socialized identity scripts that ensnare public narratives. White male heterosexuals are the unspecified norm against which alternative bodies find themselves publicly visible. Spectacularity, a hyperpublic notice, exists in direct relationship to ethnicity and gender. In the United States, this has traditionally meant that blacks and women find themselves noticeable in public ways that scan and sculpt perspectives regarding their private personhood. In this national script, the bodies of women and blacks are always and already public.[17]

Writing about the legal doctrine of privacy, the legal theorist Jed Rubenfeld argues that personhood "has so invaded privacy doctrine that it now regularly is seen either as the value underlying the right or as a synonym for the right itself."[18] This invasion creates a conundrum for Rubenfeld, whose essay "The Right of Privacy" considers the central place of sexuality in privacy cases. The social specularity attached to women and blacks explains the focus. Rubenfeld recalls in his essay the provocative notion of an "inviolable personality" and considers it as a flawed but nonetheless cognizable notion of personhood within privacy doctrine.[19] He defines personhood, the center of privacy claims, as "whatever it is that makes you *the* person you are, as opposed to whatever it is that makes you *a* person—a human being." For Rubenfeld, "*the* person" is one who operates in a social context and in the public domain. "*A* person" is the one who is inviolable—simply human. How do *the* private (social)

---

16 See Rothman, *Strangers at the Bedside*, for a history of the transformation in the American public's view of medicine, and of the influx of law and politics into the once insular domain of the physician.

17 Consider, for example, the tortured efforts on display during the Senate Judiciary Committee's hearings during the summer of 2009 regarding the nomination of Sonia Sotomayor to the Supreme Court. It was painfully difficult for some senators to separate her ethnicity from their gaze—as was most notably and egregiously on display with Senator Tom Coburn (an Oklahoma Republican), who responded to her example of what a gun law might allow and disallow with a phrase popularized during the 1950s on the television show about a Cuban entertainer, Ricky Ricardo. "You have lots of splainin' to do," Coburn jokingly responded to Judge Sotomayor—who is Puerto Rican, not Cuban.

18 Rubenfeld, "The Right of Privacy," 744.

19 See Warren and Brandeis, "The Right to Privacy."

person's acts become not merely human, but vulnerable to public notice? The link between privacy case law and an "individual's personal identity" may or may not interfere with a "personhood thesis." Rubenfeld identifies a riveting result that derives from notice: "the great peculiarity of the privacy cases is their predominant, though not exclusive, focus on sexuality ... the network of decisions and conduct relating to the conditions under which sex is permissible, the social institutions surrounding sexual relationships, and the procreative consequences of sex."[20]

Legal interest in privacy and *the* body depend on an intimate, gendered, sexual specularity. Locating the legal consideration of personhood within the law's notice of sexuality indicates that a person's private self might be considered an act (rather than a facet) of identity. This reasoning certainly explains the rulings on sodomy cases such as *Bowers v. Hardwick* and *Lawrence v. Texas*.[21] In other words, the law reaches private identity by making private acts public. The issue that makes privacy an act does not depend on the claim to an intimate and ultimately private identity. Instead it rests upon a public expression of that private identity. When Samuel Warren and Louis Brandeis worried in an 1890 essay that what is "whispered in the closets shall be proclaimed from the housetops," their concern was that the domestic shelter that protected private identities and private acts was vulnerable to public gossip.[22] What acts of identity might these be? Well, what acts of identity, despite their presumptive privacy, are most vulnerable to notice? If they are, as Rubenfeld claims, acts of sexuality, how do we understand the link between sexuality and the ways in which women become the public expression of legal notice? Is identity an act, or some body? We find the answer when we notice which bodies are dislocated from normative gendered and heterosexual privilege.[23]

Rubenfeld is correct to locate the origin of privacy claims in case law that regards (a word that bespeaks the public) expressive sexuality. Although a sense of one's self will certainly develop within the deeply personal and arguably intimate familial context, the making of a person is a complex enterprise. The success of the choices and personalities involved will depend on the manner in which they can be negotiated outside of the private. When I was a youngster, a familiar retort in playground battles—as well as from some adults who disapproved of certain conduct—was to "act your age, not your color!"[24] My sisters, playmates, and I took this admonition as a form of rebuke that we were not only being immature, but also acting in a way that made public a noticeably unflattering racial

---

20 Rubenfeld, "The Right of Privacy," 753.

21 *Bowers v. Hardwick* was a 1986 U.S. Supreme Court decision that upheld the constitutionality of a Georgia sodomy law that criminalized consenting adults' private acts of anal and oral sex. The Supreme Court directly overruled *Bowers* in 2003, in *Lawrence v. Texas*, holding that the Georgia sodomy law was unconstitutional.

22 Warren and Brandeis, "The Right to Privacy," 195.

23 Lauren Berlant writes a cultural studies version of Rubenfeld's legal argument, noting that "privacy's fall from the utopia of normal intimacy finds the law articulating its subjects as public and American through their position within a hegemonic regime of heterosexuality, which involves coordination with many other normative social positions that are racially and economically coded toward privilege" ("The Subject of True Feeling," 126). Like Rubenfeld, Berlant argues that privacy becomes a cognizable legal issue in regard to sexuality.

24 I refer to this comment in *Codes of Conduct*, 3.

stereotype that attached skin color to conduct. However, there was a positive public claim as well. When we excitedly called friends and family to tell them there was a "black person on TV," it was an acclamation that assured us that in the widest gaze possible (at that moment, television) there was also an acceptable way to be publicly colored.

But there were ethical perils that accompanied public notice—the kind that prompts sometimes voiced, sometimes silent pleas that "I hope it wasn't a black person!" when black Americans hear about a criminal act, perhaps on the radio or in the newspaper. In other words, one could ostensibly win with being publicly colored, but it was also a risk to achieve public notice. Race mattered. For identities with racial or gendered descriptors, the inherent contradiction is that each is as deeply invested in the security implicit in private selfhood as it is in a public sociality. The irony is that the public can supersede and even displace the private.

Ethics and Autonomy

In *The Ethics of Identity*, the philosopher Kwame Anthony Appiah explains that identity has its own demands, especially in the ways that "our selves are embedded in social forms."[25] Appiah notes that this consequential (public) pluralism—the social forms of self-expression—can challenge "ethical individualism." His argument is a critical notice of private and public selfhood. The ethical individualism enacted in social forms is decidedly public. Is it possible then, that privacy, identity, and ethics might find a shared location? The conjunctive association (ethics and identity) rather than the modified association evident in the title *The Ethics of Identity* is an important syntactic construction that proffers an alliance between ethics and identity. The parallel structure joins these two together and satisfies my interest in linking them to the formations of bioethics' principles. This book's interest in identity begins with its critical association with autonomy—often stated as the first of the four principles that guide traditional bioethics.

Consider autonomy's etymology. It comes from the Greek *autos* (self) and *nomos* (rule). Recalling this etymology explains the importance of a focus on the identitarian (self) nature of authority. Authority, identity, and autonomy are all critical anchors to the discussion in this book. Authority stems from the Latin *auctoritas*—power that is legitimate. An associative argument for a synergistic relationship between autonomy, authority, and individualism encourages a consideration of these as relational terms that see individual liberty and agency as the authoritative (legitimate) construction of the self. Liberty and agency lie at the root of private personhood.[26]

---

25 Appiah, *The Ethics of Identity*, 73.

26 In *The Ethics of Identity*, Appiah does indeed move directly to a consideration of autonomy following his discussion of "ethical individualism" in chapter 3, "The Demands of Identity." The difference I suggest here, in associating the ethics of identity with private autonomy, might be more a directional than a substantive difference from Appiah's argument.

I am a black woman. The plural identities that I claim lie not only in my authority to do so, a private authority that adheres to a selfhood. Because of the evident visibility of the two particular categories my body occupies—gender and race—I actually have little social choice in the matter. My private identities are always and already public.[27]

When value and identity are publicly assigned and associated, it is very often because gender and race are at issue. When identity exceeds the private (that is, when we notice women and blacks essentially because they are not male and white), our notice and the visibility of these categories become a matter of ethical focus.[28] Instead of individual conduct being the initial value of difference, identity produces differential treatment before anything whatsoever is known about individual character. This is especially the case in legal and medical ethics, when the events that are so publicly rehearsed, like the Schiavo case or the utility of Henrietta Lacks's cellular matter, happen in part because the narrative scripts attached to women and blacks become public narratives that very nearly dismiss the private.

Recall the narratives that opened this discussion. Of the "right to die" cases, Terri Schiavo's extraordinarily public and riveting narrative depended on her fitting into the stereotypical categories most often associated with women. She was portrayed as a daughter and a wife in the many retellings of her story. Her tragic situation incorporated all of the historic versions of female helplessness that social histories rehearse. Quite literally, Schiavo could not speak, think, or act on her own behalf. Although these are extreme versions of bias, she fit neatly into those extremes as well as the narratives that we have traditionally attached to the female gender.

Consider, for example, which among the memorable right to die cases are about men. Nancy Cruzan, Terri Schiavo, Elizabeth Bouvia, and Karen Ann Quinlan—the most memorable names associated with these cases are all female. It is certainly true that men have faced the same medical dilemma. But it is the stories of women that make it from the private to public consumption. In the Schiavo case, among those who testified that they knew what Ms. Schiavo would have wanted was her sister-in-law. William Levesque covered her testimony for the *St. Petersburg Times*:

> Joan and Terri Schiavo had become close friends in the mid 1980s. Joan Schiavo
> said she told her sister-in-law about a friend who was forced to end life support to

---

27 The potential public exposure resident in these categories was particularly salient at the historical moment in the United States when political attention focused on Barack Obama as a black man and Hillary Clinton as a woman—both candidates for the Democratic Party's presidential nomination. Notice, however, that whiteness was erased for Clinton in the mainstream media's public discourse surrounding her candidacy—her race was not a category of spectacle; only her gender was the visible mark.

28 Of course there are times of ambiguity when race and women's gender are not evident. The argument about the spectacularity of these categories nevertheless holds, because these occasions are likely to spark even more public notice and queries about "what are you?" The biography of Anatole Broyard, former *New York Times Book Review* editor, attracted a good deal more attention than it might have when it was disclosed that Broyard, who had lived his professional and personal life as a white man, was revealed to be black. See Gates, "The Passing of Anatole Broyard." Broyard's daughter wrote a memoir following her father's death that explores the history of her family and her own discovery and engagement with her mixed-race background (Broyard, *One Drop*).

an infant after health problems. Terri Schiavo told her that she would have done the same thing for the baby if its life could not otherwise be saved, Joan Schiavo testified. Joan Schiavo also heard other comments after she and Mrs. Schiavo saw a movie about someone who had an accident and was in a coma. "We had stated that if that ever happened to one of us, in our lifetime, we would not want to go through that. That we would want it stated in our will we would want the tubes and everything taken out. She did not like the movie. Just the whole aspect of family and friends having to come and see their son or friend like that, she thought it was horrible."[29]

Neither Terri Schiavo's own voice nor that of her sister-in-law mattered enough to make state or national legislators honor her wish. The legal spectacle that attached to Schiavo's life was relentlessly gendered in the ways that social context has delivered more authority to men's voices than to women's. The collapse of Schiavo's private identity, nearly obscured in the media circus that surrounded the event of her dying, is a reasonable location for ethical review. The ethics of this matter cannot dismiss the gendered politics in the public space her body came to occupy.

Bibliography

Appiah, Kwame Anthony. *The Ethics of Identity*. Princeton: Princeton University Press, 2005.

Berlant, Lauren. "The Subject of True Feeling: Pain, Privacy, and Politics." In *Left Legalism/Left Critique*, edited by Wendy Brown and Janet Halley, 105–33. Durham: Duke University Press, 2002.

Broyard, Bliss. *One Drop: My Father's Hidden Life; A Story of Race and Family Secrets*. New York: Little, Brown, 2007.

Charon, Rita. *Narrative Medicine: Honoring the Stories of Illness*. Oxford: Oxford University Press, 2006.

Charon, Rita, and Martha Montello, eds. *Stories That Matter: The Role of Stories in Medical Ethics*. New York: Routledge, 2002.

Gates, Henry Louis, Jr. "The Passing of Anatole Broyard." Reprinted in Henry Louis Gates, Jr., *Thirteen Ways of Looking at a Black Man*, 180–214. New York: Random House, 1997.

Gold, Michael. *A Conspiracy of Cells: One Woman's Immortal Legacy and the Medical Scandal It Caused*. Albany: State University of New York Press, 1986.

Harris, Trudier. "This Disease Called Strength." *Literature and Medicine* 14 (1995): 109–26.

Holloway, Karla FC. *Codes of Conduct: Race, Ethics, and the Color of Our Character*. New Brunswick, N.J.: Rutgers University Press, 1995.

---

29 William R. Levesque, "Schiavo's Wishes Recalled in Records," *St. Petersburg Times*, November 8, 2003 (http://www.sptimes.com).

Landecker, Hannah. *Culturing Life: How Cells Become Technologies*. Cambridge: Harvard University Press, 2007.

Nelson, Hilde Lindemann, ed. *Stories and Their Limits: Narrative Approaches to Bioethics*. New York: Routledge, 1997.

Rao, Radhika. "Genes and Spleens: Property, Contract, or Privacy Rights in the Human Body?" *Journal of Law, Medicine and Ethics* 35, no. 3 (fall 2007): 371–82.

Rothman, David. *Strangers at the Bedside*. New York: Basic, 1992.

Rubenfeld, Jed. "The Right of Privacy." *Harvard Law Review* 102, no. 4 (February 1989): 737–807.

Skloot, Rebecca. "Henrietta's Dance." *Johns Hopkins Magazine*, April 2000 (http://magazine.jhu.edu/).

———. *The Immortal Life of Henrietta Lacks*. New York: Crown, 2010.

Warren, Samuel, and Louis Brandeis. "The Right to Privacy." *Harvard Law Review* 4, no. 5 (December 15, 1890): 193–220.

Legal Cases

Bowers v. Hardwick, 478 U.S. 186 (1986)

Lawrence v. Texas, 539 U.S. 558 (2003)

Attending to Difference

Phenomenology and Bioethics

Philipa Rothfield

....

Although it has never dominated traditional bioethics, phenomenology has been quietly generating critical perspectives on medicine, while offering its own conceptual alternatives. Phenomenology has been at pains to establish that the patient's experience of illness is ethically paramount, and to suggest ways in which this recognition might be incorporated within medical practice. Its elucidation of the patient's experience within the biomedical setting is typically conducted at an abstract, formal level. According to such investigations, the particular merely serves to facilitate an understanding of the essential workings of subjectivity.

Judith Butler has criticized Merleau-Ponty, both for failing to specify the kinds of bodies and sexualities he was phenomenologically analyzing and for the unacknowledged intrusion of his own sexually specific understanding of the matter (Butler 1989). Butler's charge was that Merleau-Ponty did not consider *whose* bodies and *which* sexualities were at stake, nor did he acknowledge his own corporeal complicity in the way in which he viewed the subject. Thus, despite his intention to describe certain moments in the general structure of human, sexual being, Butler alleges that Merleau-Ponty's analysis was both partial and skewed.

Is it possible that phenomenological bioethics is similarly skewed? Should it too specify the kinds of bodies that it takes to participate within biomedical practice? And if it does, will this impact upon the character of its claims? To ask these questions is to critically review phenomenological bioethics in the light of postmodern sensitivities regarding difference. Phenomenology has a lot to offer in that it provides a philosophical commitment to everyday life as it is lived in corporeal terms. Yet, if difference is missing from the account, how is it possible to discern the diversity of bodily experience? Is it possible to generate a phenomenological bioethics that retains its formal validity while engaging with the differences that exist between bodies? This chapter pursues these questions by looking at cultural differences among bodies within medicine. It aims to ask of phenomenological bioethics whether it can deal with cultural differences as they occur within the medical setting.

Beginning with a brief account of phenomenology's approach to medical ethics, the challenges of cultural difference are introduced, then explored in relation to questions of corporeal specificity. Although there is little material on cultural difference in relation to phenomenological bioethics, medical anthropology looks at medical practice in a range of cultural settings. Set within this context, Thomas Csordas's work confronts the epistemological complexities of intercorporeal communication and interaction. Using a phenomenological framework of analysis, he highlights the problems that attend the effort to understand the corporeal specificity of another when that other is culturally different. Phenomenologists such as Zaner and Toombs recognize that the doctor-patient relation is not symmetrical. However, they do not dwell upon the effect that cultural difference has on this relationship. Their frameworks do not specify what sorts of bodies enter into the biomedical exchange. Csordas's work invites a response from phenomenology in regard to the bioethical dimensions of intercorporeal difference.

Phenomenological Bioethics

Although each approach has its nuances, phenomenology characteristically aims to analyze the structures of experience. In the case of biomedicine, it is the patient's experience of illness that is examined. Phenomenologists variously elucidate the impact of illness or injury on the subject, including the disruptions of everyday life, the experience of symptoms or discomfort, alien body sensations, or bodily distortions; in short, the existential dimensions of illness. They argue that doctors are not necessarily trained to recognize the centrality of the patient's experience in their work because the scientific paradigm tends toward a reductionist view of the person (Zaner 1988). Many phenomenologists trace the reductive tendencies of medicine to Descartes' mechanistic reading of the body. According to Leder:

> At the core of modern medical practice is the Cartesian revelation: the living body can be treated as essentially no different from a machine. Though any good clinician also engages the patient-*as-person*, the predominant thrust of modern medical therapeutics has been upon such mechanistic interventions. (Leder 1992: 23)

In contrast, phenomenological accounts of human subjectivity, especially arising from the work of Merleau-Ponty, understand the self as inextricably mind and body, thereby resisting both dualism and reductionism.

Phenomenology's remedy for the reductions of medicine is to propose a focus for medical practice that incorporates a closer attention to the vicissitudes of corporeal life, especially the patient's. Leder (1992) refers to the lived body and the existential dimensions of the medical setting, whereas Toombs

(1992) emphasizes immediate experience and the meaning of illness. Shanner (1996) likewise draws attention to the experience of those patients who have lived through certain predicaments in order to stress what it was like for them. Both Zaner and Shanner refer to patient narratives as a means to understand and represent people's experiences. The thrust of these approaches is to make space for a fuller disclosure of the patient's experience of illness. Their aim is to highlight the importance and complexity of experience within medicine.

Given the divergence between the allegedly reductive views of bio-medicine and the existential domain of the patient, it is not surprising that the medical encounter may fail to achieve a mutual understanding between the patient and the doctor. Toombs claims that there is a systematic distortion of meaning in the physician–patient relationship:

> [I]llness is experienced in significantly different ways by physician and patient. Consequently, rather than representing a shared "reality" between them, illness represents in effect two quite distinct "realities." (Toombs 1992: 10)

At present, it is the patient's experience of illness that is elided in the medical encounter, for the discourse of medicine prevails over the experiential lifeworld of the patient. This is understandable given the sense of therapeutic intervention that lies at the heart of medical practice. However, Toombs argues that physicians need to attend to the lived experiences of patients in order to set therapeutic goals (Toombs, 1992: xvi). She writes:

> As a first step towards achieving a shared understanding of the meaning of illness, physicians can learn to recognize and pay attention to, these typical characteristics of the human experience of illness. (Toombs 1992: 97)

Baron argues for a similar shift on the part of the doctor, toward an engagement with the patient's needs, desires, and demands (Baron 1992: 44).

This requires the construction of a shared world of meaning between doctor and patient, one that incorporates the patient's subjective experience of illness, a side of the equation systematically undervalued in relation to the scientific attitude embedded in the medical gaze. Despite structural differences between doctor and patient, both Baron and Toombs aim for a convergence of understanding within clinical practice. Such an understanding, they argue, needs to incorporate the experiential standpoint of the patient. Given that the scientific perspective of the physician does not conventionally incorporate patient experience, as distinct from patient diagnosis, the sort of epistemic shift required by the physician is significant.

How, then, is such a shift to be achieved? Toombs approaches the issue on the basis of that which is common among people—their lived corporeality. In her view, everyday experience can function as a basis for comprehending the disruptions of illness. One aspect of such everyday experience is the ambiguity of the lived body. According to Toombs, everyone has a sense of the ambiguity of lived corporeality: that I am my body but also separate from it (it is outside my control). I realize that I am my body because all my efforts involve physical activity, yet I realize that I cannot always control my body; I tire, I get sore, I get indigestion, I miss that backhand shot. For Toombs, illness confronts us with this not-unfamiliar state of affairs—on the one hand, I am this sick body, and on the other, its being ill is out of my control. There is also for Toombs a sense in which everyone experiences their bodies as alien; this is said to be the sense in which we all feel our bodies in terms of their limitations, as encumbrances.

All people, then, experience bodily ambiguity and the alien character of corporeality, although the experience may be more extreme in illness. These experiential facts are taken to indicate that the "lifeworlds of physician and patient do indeed provide for mutual understanding with regard to the illness experience" (Toombs 1992: 98). The lifeworld itself also enables the physician to understand particular bodily disorders (Toombs 1992). Here, Toombs cites Engel and Engelhardt, who claim that the development of a scientific understanding of illness is built upon a more personal sense of one's own corporeal life processes. First, medical students draw upon their own bodily experience in order to develop a scientific understanding of bodily processes, and second, doctors relate the patient's experiences to elements of their own experience.

Cassell's work, written for doctors, also suggests that the lived body is a means toward effective communication:

> Because all of us have bowel movements, we all have a framework of reference to help us understand her. Persons taking histories should use themselves and their own experiences with their bodies and the world as a reference for what they hear. (Cassell 1985: 46)

In cases where the doctor has never had a particular experience, Cassell suggests that more questions need to be asked "so that when you are finished you have both acquired the diagnostic information and learned more about the world" (Cassell 1985: 46). There is a sense here that the gap between one corporeality and another is not that great.

In sum, Toombs argues that the humanity of physician and patient is a means to bridge perspectival, discursive, and lifeworld differences between the two parties. By emphasizing the utmost generalities of corporeal subjectivity, she asserts a common basis for the development of mutual understanding. Both Toombs and Cassell suggest that the doctor draw upon his or her own bodily experience and understanding in order to comprehend the experiential perspective of the patient. Even where that experience is lacking, it is thought that connections can be made. For Toombs, it may be possible to extend certain experiences of wellness into the domain of illness, while for Cassell, questioning can expand the doctor's understanding. In both cases, the doctor is potentially capable of reaching a corporeal understanding of the patient's lived body, one that incorporates elements from his or her own lived body.

Is it, could it, ever be the case that differences between the doctor and the patient stand in the way of that corporeal understanding? Rather than begin with universalist assumptions about human corporeality, what if we began with the difficulties that bodily specificity brings to the bioethical situation? What might this suggest to the project of phenomenological bioethics? In what follows, a number of cases have been selected that challenge the notion that a common corporeality underlies all parties to the medical encounter. These examples illustrate the ways in which racism and ethnocentrism in particular both enter into and trouble the biomedical scene.

Differential Treatment on Grounds of Color

Case 1

Although Cassell nowhere mentions skin color in his work on clinical technique for doctors, perceptions of color may well influence the way in which doctors approach their patients and vice versa. In her paper, "Reconstructing the Patient, Starting with Women of Color," Dorothy Roberts (1996) argues that doctors treat women of color differently than their white female patients. She cites a number of studies which have found that women of color get worse treatment, less explanation, wait longer to be treated, and are told, not asked, what to do (Roberts 1996). Roberts claims that because of these negative experiences, these women are now thoroughly alienated from the medical establishment. As a consequence, reforming the ethics of the medical profession may not be enough to reverse the trend of poor treatment. The point is that it is not enough for white doctors to finally come to the party (were they to address their racism), for these women are skeptical, critical, and resistant to what they perceive to be "medical control." Roberts concludes by asking whether some other strategy of empowerment for these women might not be more effective, one where they have more control over the institutions that deliver health care (Roberts 1996: 136).[1]

Case 2

This example concerns the delivery of palliative care to the Pitjantjatjara people of central Australia. According to Jon Willis:

---

1 The issue of control, and therefore empowerment, appears quite central. Sir Gustav Nossall recently commented on national radio (3RN, September 5, 2001) that improvements in Australia with regard to indigenous health care have been significantly greater in local, small clinics run and largely staffed by indigenous people. Roberts similarly refers to the National Black Women's Health Project, which aims to facilitate black women taking control over their own health. In policy terms this is to recognize that "a feminist reconstruction of the patient should not encompass solely the physician's understanding of the patient, but also the patient's understanding of the physician" (Roberts 1996: 136).

[T]he issue is not simply to modify elements of palliative care so that cultural differences in belief and practice are accommodated, but to recognise that different cultures "do death" in different ways, and that institutions for the provision of palliative care are bound up in the "way of dying" of the culture in which they originated. (Willis 1999: 427)

Willis argues that it is not possible to provide for what he calls an "acceptable death," unless that care recognizes and accommodates the specific "way of dying" of Australian Pitjantjatjara Aborigines as a cultural aspect of these peoples' lives. He describes two elements of Pitjantjatjara culture that especially impact upon their dying. One is the importance of land, which relates to ancestral and dreamtime associations with land, and the second is the importance of being cared for by the mother's family rather than health care professionals. Given these factors, it could be argued that state-funded palliative care ought to reflect the Pitjantjatjara people's need to die in their homeland, under family care. Unfortunately, this contrasts with the fact that most palliative care in Australia is hospice or hospital-based, and that "patients in rural settings have typically been required to relocate to regional centres or even state capitals to access any structured form of palliation" (Willis 1999: 342). Furthermore, the costs associated with providing care to extremely remote locations would be high, taxing the scarce resources currently available in terms of district nursing staff. On the other hand, if patients are forced to relocate, this can provoke fear and isolation (Willis 1999: 425). A study of the experiences of indigenous patients on center-based dialysis for end-stage renal failure revealed the great social and cultural costs involved in removing patients from their homes in central Australia (Willis 1995). Moreover, once they were relocated, it was noted that race issues affected patient–carer interactions, a finding echoed by Preece, who cites one patient's remark:

On average I take between 100 and 260 tablets a week. Pathology tests are performed monthly and the results are never discussed with me. Why am I having these tests done, and for what? There is no feedback from any of the hospital staff. ... I feel the doctor is intimidated by me because I speak up if I am not happy with the service provision. I am also a spokesperson for the ATSI [Aboriginal and Torres Strait Islander] patients here on dialysis. *Sometimes I think the doctor wishes I would die so I would not be a problem.* (Preece, cited in Willis 1999: 428; my emphasis)

[...].

Corporeal Spectatorship and the Cultural Other

Ethnographer and cultural phenomenologist, Thomas Csordas, focuses upon the means by which one bodily subjectivity, typically that of the ethnographer, might understand and apprehend another, typically that of a cultural other. Like his fellow phenomenologists, Csordas writes of the lived body and embodiment as the field within which other subjectivities are articulated. But—and here lies the difference—as an ethnographer, he recognizes the cultural character of lived corporeality. Csordas takes the field of embodiment to signify the inextricability of culture and experience. Cultural phenomenology is likewise supposed to synthesize "the immediacy of embodied experience with the multiplicity of cultural meaning in which we are always and inevitably immersed" (Csordas 1999: 143). The point is that the field of "immediate" experience is neither prior to nor outside of culture, but is articulated within and according to a heterogeneity of cultural meanings and practices. To this extent, any understanding of self and body is necessarily "unstable and culturally variable" (Csordas 1999: 143). There is no universal, invariant bodily self. As Victora has put it, "people know their bodily facts in different ways" (Victora 1997: 170).

The recognition that bodily subjects are always culturally specific does not merely apply to the cultural other as an object of knowledge but also extends to the knowing subject. It pertains to the corporeal means by which the knowing subject apprehends the specificity of the other, via what Csordas calls the exercise of somatic attention. Somatic attention is characterized by Csordas as those "culturally elaborated ways of attending to and with one's body in surroundings that include the embodied presence of others" (Csordas 1993: 138). There are two salient features of somatic attention. First, we attend with or through the body, and second, this mode of attention is culturally and socially informed. In other words, "neither attending to nor attending with the body can be taken for granted but must be formulated as culturally constituted somatic modes of attention" (Csordas 1993: 140).

In a sense, we are returned to Cassell and Toombs—the doctor understands the patient's condition by reference to his or her own corporeality. However, the difference lies in the fact that for Csordas, having a culturally specific body means also that one perceives and understands the world in a culturally specific manner. I understand what is happening in my own and other's bodies through my own body. The culturally specific body is not just a cultural artefact, it is our means of experiencing the world. Csordas's point is that experience itself involves the exercise of certain kinds of sense and sensibility, and that these embodied faculties are permeated by culture and cultural values (Csordas 1999: 155). The concept of somatic attention as a cultural, corporeal mode of apprehension signals lived corporeality as the manner by which one person engages with another. The tendency for phenomenological bioethics to emphasize the universal aspects of experience renders the cultural specificity of the body epiphenomenal, whereas Csordas sees the body (embodied subjectivity) as the lynchpin of intersubjective communication and apprehension.

To return to the medical setting, and in light of Csordas's analysis, we might say that the doctor's comprehension of the existential situation of his or her patient occurs according to an act of corporeal spectator-ship. Where the doctor hails from a different cultural setting than his or her patient, he or she is somewhat like Csordas's ethnographer attempting to understand a form of cultural alterity. By way of comparison, the doctor, like the ethnographer, can only comprehend the patient through the specificity of his or her own lived corporeality. The ethical difficulty is that the doctor's specificity could well act as a filter for the experiential reality of the cultural other. In such a case, difference might be effaced, distorted, or disregarded in favor of some projection on the doctor's part. Perhaps the doctor has read a pile of cultural factfiles and now feels qualified to make culturally sensitive diagnoses. Csordas refers to an ethnographer's adapting to the everyday practices of a Nepalese community in order to understand the bodily metaphors inscribed within its spiritual and religious rituals (Csordas 1999: 155). This is the point at which the analogy between doctor and ethnographer fails, for the doctor is not going to go anywhere in order to be able to apprehend the corporeality of a cultural other. Rather than adapt to the patient, there is a greater likelihood that the doctor will impose the familiar, where otherness is misrecognized as sameness. Remember the hospice worker in Gunaratnam's study, who was worried that she might only respect those cultural attitudes that did not impinge on her own values (Gunaratnam 1997: 178)? Here the question has to do with whether cultural and corporeal values are perceived at all, and whether they are misrecognized as like one's own values or, alternatively, are taken to be foreign to the point of alienation (as in the exoticism of Orientalism, for example). It also concerns the invisibility of whiteness; who "oneself" is. If a "woman of color" is cited as a possible other, who is she other to? If she is perceived as the same, who functions as the norm of sameness?

It is in light of these difficulties that we might consider Smaje and Field's assertion that some thought should be given to employing people with ethnic backgrounds similar to those of patients:

> One important step might be to encourage greater recruitment of palliative
> care staff from minority ethnic groups. No data are available on current levels
> of employment, but certainly hospices are often regarded as fairly "white"
> institutions. (Smaje and Field 1997: 159)

We might likewise consider Roberts and Nossall's point regarding empowering people to set up and run their own community health care centers. Multicultural medical clinics conceived along similar lines would have to deal with the signature ethnicities of their community, instituting mechanisms by which client needs could be recognized and accommodated.

The clinical setting contributes to the kind of sensibility that is deployed in the medical encounter. Roberts depicts this in regard to the public hospital settings where women of color get a good dose of racism along with their medical treatment. The point is that institutional protocols condition the subjectivity and sensibility of its professionals. The somatic attentions of a hospice worker in multi-cultural Britain or Aboriginal Australia derive in part from the values enshrined in the management of death. Although not apparent, management protocols are liable to be predicated upon particular cultural and, no doubt, other forms of dominance. Thus we encounter the indigenous Australian patient who feels his doctor sometimes wishes he would die, and the "noisy" Greek and North

African mourners in the English hospice. The perception of these sick and grieving peoples is based on the deployment of a certain sensibility that ensues in part from the cultural values embedded in institutional sites and discourses.

Csordas's work on somatic attention produces the insight that the doctor–patient exchange is always refracted through the cultural and corporeal sensibilities of its participants. While Leder, Baron, Toombs, and others lead us into the clinical encounter, and encourage us to value the patient's experience, they do not see corporeal differences as impacting upon the doctor's means of understanding, or the phenomenologist's for that matter. Cultural difference is envisaged as having something to do with the patient, but not particularly the doctor. Although Toombs admits that "cultural meanings are an important determinant in the manner in which illness is apprehended" (Toombs 1992: 36), these meanings are taken to form an interpretative layer that coats a prereflective universality of experience. She thus writes of the meanings assigned to bodily sensations as if the sensation itself has an integrity that preexists its cultural interpretation. Similarly, pain is taken to be a universal experience whose significance is all that varies from culture to culture. Toombs's solution to the problem of cultural difference is, like Cassell's, for the doctor "to ask the patient" in order to reveal the underlying experience that is common to all (Toombs 1992: 45).

Cultural difference for such theorists is information, whereas for Csordas it goes to the heart of intercorporeal communication, implicating both doctor and patient. Inasmuch as the doctor endeavors to comprehend the patient's predicament, his or her corporeality is brought to the fore, for it is the means by which the other can be understood. The doctor's attempt to understand is an act of corporeal spectatorship. Although not emphasized in Csordas's framework, corporeal spectatorship occurs within a discursive and institutional setting. The protocols of these settings both embrace cultural values and authorize the use of perceptual sensibilities that embody these values. In the medical setting, this means that the doctor might well impose certain norms in making perceptual and diagnostic judgments. It also means that the apprehension of people's behavior reflects as much upon the perceiver and the site from which the perception ensues as it does upon its object.

The obstacle medicine faces that ethnography does not is its presumed universalism. Phenomenological bioethics at its best draws our attention to the concrete lived bodies that participate in the medical encounter. It also speaks to the tendency for the doctor to diagnose rather than apprehend the lived corporeality of the patient. Having transported us to the space inhabited by the doctor and patient, however, Csordas alerts us to the cultural specificity of corporeal spectatorship. Henceforth it becomes impossible to give a culturally neutral reading of the somatic interchange. The doctor's ability to perceive the patient is formed according to a number of factors relating to his or her cultural milieu.

Were phenomenological bioethics to address this state of affairs, it would concern itself with the kinds of ethical complications that arise because of such differences between doctor and patient. These include the degree to which cultural factors are also embedded in the institutional settings in which doctor and patient are found. The issue is not merely one of difference but concerns the relative status of the players and the prevalence of certain values over others. It also concerns the relations of power embedded in the physician's subject-position and professional setting, and relates in turn to the patient's perception of the doctor.

To speak of cultural difference is often to identify the cultural other without identifying the one who is culturally the same, that is, whose privilege renders their cultural status invisible. Aileen Moreton-Robinson has shown the many ways in which whiteness remains invisible within Australian feminism, despite the attempt to represent difference (Moreton-Robinson 2000). The point is that white privilege allows whiteness to remain unmarked, even where color and cultural difference are on the table:

> The representation of "difference" in feminism was first articulated as gender, culture and class differences between white men and women. ... As the foregoing literature review shows, whiteness as subject position, "race," privilege and dominance are not marked as a difference in the early literature. ... Nor is whiteness made visible in the later literature, which focuses on differences between white women and women who are "Other." (Moreton-Robinson 2000: 69)

By highlighting the cultural constitution of the corporeal spectator—the one who deploys a very particular sensibility in exercising somatic attention—Csordas reveals one of the means by which racism and ethnocentrism occur, for the bodily experience of the doctor is the perceptual filter by which the situation of the patient is apprehended. Although Csordas writes that in principle neither party has any a priori rights to objectivity, in practice, the doctor's perception and his or her institutional norms have historically determined what counts as objectivity.

Csordas has depicted the ethnographer's attempt to adapt to the culture of the other in order to understand its rituals (Csordas 1999). This strategy is not available however to the doctor. Moreover, as Moreton-Robinson points out:

> Indigenous women as embodiments of racial difference can never know what it is like to experience the world as a white woman, just as white women can never know what it is like to experience the world as an Indigenous woman. *To know* an Indigenous constructed social world you must experience it from within; to *know about* such a world means you are imposing a conceptual framework from the outside. These two ways of knowing inform us that there are limits to knowing an "Other" be they black or white and these restrictions impact on intersubjective relations and the exercising of power. (Moreton-Robinson 2000: 185)

These remarks raise questions about the possibility of intercorporeal knowledge and understanding where the parties embody very different sensibilities. To "know about" another's subjectivity is seen to be conceptual rather than experiential. If the conceptual knowledge of difference has its limits, what ethical considerations follow? Power is a crucial term here; ignorance is one thing, socially sanctioned ignorance another, and imposition another. Given the limits to "knowing an 'Other,'" perhaps empowerment and control over health care arise as necessary strategies by which to address cultural difference and racism, in that the apprehending sensibilities of the institution would reflect those of its clientele.

References

Baron, Richard (1992) "Why Aren't More Doctors Phenomenologists?" in *The Body in Medical Thought and Practice,* Drew Leder (ed.) Dordrecht: Kluwer Academic, pp. 37–47.

Butler, Judith (1989) "Sexual Ideology and Phenomenological Description, A Feminist Critique of Merleau-Ponty's *Phenomenology of Perception,*" in *The Thinking Muse, Feminism and Modern French Philosophy,* Jeffner Allen and Iris Marion Young (eds.). Indianapolis: Indiana University Press, pp. 85–99.

Cassell, Eric (1985) *Talking with Patients,* vol. 2. Cambridge, Mass. MIT Press.

Csordas, Thomas (1993) "Somatic Modes of Attention," *Cultural Anthropology* 8, 135–156.

Csordas, Thomas (1999) "Embodiment and Cultural Phenomenology," in *Perspectives on Embodiment, The Intersections of Nature and Culture,* Gail Weiss and Honi Fern Haber (eds.) New York and London: Routledge, pp. 143–162.

Gunaratnam, Yasmin (1997) "Culture Is Not Enough, A Critique of Multi-Culturalism in Palliative Care," in *Death, Gender and Ethnicity,* David Field, Jenny Hockey, and Neil Small (eds.) London and New York: Routledge, pp. 166–185.

Leder, Drew (1992) "A Tale of Two Bodies, The Cartesian Corpse and the Lived Body," in *The Body in Medical Thought and Practice,* Drew Leder (ed.) Dordrecht: Kluwer Academic, pp. 17–35.

Moreton-Robinson, Aileen (2000) *Talkin' Up to the White Woman: Indigenous Women and Feminism.* Brisbane, Australia: Queensland University Press.

Roberts, Dorothy (1996) "Reconstructing the Patient: Starting with Women of Color," in *Feminism and Bioethics, Beyond Reproduction,* Susan Wolf (ed.) New York and Oxford: Oxford University Press, pp. 116–143.

Shanner, Laura (1996) "Bioethics through the Back Door: Phenomenology, Narratives and Insights into Infertility," in *Philosophical Perspectives on Bioethics,* L. W. Sumner and J. Boyle (eds.) Toronto: University of Toronto Press, pp. 115–142.

Smaje, Chris and Field, David (1997) "Absent Minorities? Ethnicity and the Use of Palliative Care Services," in *Death, Gender and Ethnicity,* David Field, Jenny Hockey, and Neil Small (eds.) London and New York: Routledge, pp. 142–165.

Toombs, Kay (1992) *The Meaning of Illness, A Phenomenological Account of the Different Perspectives of Physician and Patient.* Dordrecht: Kluwer Academic.

Victora, Ceres (1997) "Inside the Mother's Body, Pregnancy and the 'Emic' Organ 'the Body's Mother,'" *Curare* 12, 169–175.

Willis, Jon (1995) "Fatal Attraction, Do High Technology Treatments for End-Stage Renal Disease Benefit Aboriginal Patients in Central Australia?" *Australian Journal of Public Health* 19(6), 603–609.

Willis, Jon (1999) "Dying in Country, Implications of Culture in the Delivery of Palliative Care in Indigenous Australian Communities," *Anthropology and Medicine* 6(3), 423–435.

Zaner, Richard (1988) *Ethics and the Clinical Encounter.* Englewood Cliffs, N.J.: Prentice-Hall.

Toxic Life in the Anthropocene

Margaret Lock

The Embodiment of Trauma

An article published in the mid-1990s by the epigeneticist Michael Meaney has become iconic of the field of behavioral epigenetics. This research made use of a model of maternal deprivation created in rats by removing young pups from their mothers shortly after birth, thus terminating maternal licking and grooming crucial to their development. The deprivation altered the expression of genes that regulate behavioral and endocrine responses to stress, and hence, indirectly, hippocampal synaptic development. It was found that these changes could be reversed if pups were returned in a matter of days to their mothers. Furthermore, when the birth mother was a poor nurturer, placement of her deprived pups with a surrogate mother who licked and groomed them enabled the pups to flourish. Crucially, it was shown that pups or foster pups left to mature with low-licking mothers not only exhibited a chronically increased stress response but also passed on this heightened sensitivity to stress to their own pups. Hence, variation in maternal behavior results in biological pathways causal of significantly different infant phenotypes that can persist into adulthood and are potentially transmitted to the next generation.

A substantial body of work based largely on animal models, but increasingly in humans, substantiates these findings and broadens their significance. Beginning in the 1990s a literature has accrued showing a strong relationship between "childhood maltreatment" and negative mental health outcomes ranging from aggressive and violent behavior to suicide. Current investigations are gradually exposing how the "biological embedding" of childhood maltreatment comes about. The overall conclusion drawn from this research is that the "epigenome" is responsive to developmental, physiological, and environmental cues that can result in so-called "epigenetic marks."

In 2011 Moshe Szyf titled a presentation he gave at a Montréal gathering "DNA Methylation: A Molecular Link between Nurture and Nature." At the time this talk was given, evidence for such a link had accrued primarily from animal research and from one human study based on a sample of twenty-five individuals who had suffered severe abuse as children and later committed suicide. At autopsy, the donated brains of these individuals showed a significantly different pattern of DNA methylation than did those of a control group of sixteen "normal" individuals. A second control group of twenty individuals who had committed suicide but where no known abuse had taken place was also included in the study. The findings are presumed to substantiate a mechanism whereby nature and nurture meld as one. In this particular case, childhood adversity is associated with sustained modifications in DNA methylation across the genome, among which are epigenetic alterations in hippocampal neurons that may well interfere with processes of neuroplasticity.

The researchers acknowledged that the sample was small and that the study cannot be validated. The absence of a control group that experienced early life abuse and did not die by suicide is another shortcoming. Furthermore, the abuse that the subjects experienced was exceptionally severe. Szyf and colleagues readily agree that deep understanding of these processes is rudimentary. Even so, given that epigenetic markers have been shown to play important roles in learning and memory that may be transmitted intergenerationally, these findings suggest how the trauma associated with events such as colonization, slavery, war, and displacement may be transmitted through time. It is possible that they may also bring about insights in connection with resilience to such events.

Sculpting the Genome

The epigenetic mechanism best researched to date is methylation, a process in which methyl groups are added to a DNA molecule. DNA methylation is found in all vertebrates, plants, and many non-vertebrates. Enzymes initiate such modifications that do not alter the actual DNA sequence, but simply attach a methyl group to residues of the nucleotide cytosine, thus rendering that portion of DNA inactive. Epigenetic researchers are careful to point out that the identification of mechanisms that transmit signals from social environments external and internal to the body resulting in DNA methylation have yet to be fully worked out. But it is incontrovertibly demonstrated that methylation functions so that any given genome is able to code for diversely stable phenotypes. In other words, although every cell at the time of formation is "pluripotent," that is, has the potential to become any kind of mature cell, methylation brings about so-called "cell differentiation," resulting in, for example, liver, neuronal, or skin cells. Furthermore, such changes take place not only in utero and early postpartum years, but continue throughout the life span.

Relatively recently it has been recognized that environmental exposures originating outside the body bring about changes to the three-dimensional chromatin fiber that compacts DNA inside cells. The idea that the epigenome as a distinct layer over and enveloping the genome, is no longer acceptable. Genome and epigenome are now recognized as a single flexible, commingled entity, orchestrated by shape-shifting chromatin that at times results in changes that are hereditable. In addition, small strips of DNA can be altered, often during replication, some of which changes result in mutations that may or may not be hereditable. Epigenetic mechanisms other than methylation also regulate gene expression, but they have yet to be researched to the same extent as has methylation.

In sum, genes are "catalysts" rather than "codes" for development, and it is the structure of information rather than information itself that is transmitted. DNA is not changed *directly* by environmental exposures. Rather, whole genomes respond ceaselessly to a wide range of environments, exposures, and experiences, and chromatin mediates such responses that, in turn, modulate DNA expression. The methylation processes described above are manifest in several time scales—evolution; transgenerational inheritance; individual lifetimes; life-course transitions (including infancy, adolescence, menopause, and old age) and seasonal change modifications. These effects, miniaturized in individual bodies, are researchable at the molecular level.

In the remainder of this essay I highlight the impact of toxins on the physical body, toxins which we all imbibe daily and are exposed to in ever-increasing amounts. Of course, certain populations and individuals are pervasively impacted much more than are others. Accounts given by affected individuals about toxic exposures are in part informed by sensations experienced in the material body itself. Such sensations are necessarily contingent on evolutionary, historical, environmental, economic, political, and social variables that everywhere contribute to the state of individual bodies. Such sensations result from direct exposures throughout life, and/or from attributes inherited intergenerationally.

Embodiment is further constituted by the manner in which self and others represent the body, drawing on local categories of knowledge and experience, including scientific knowledge to the extent that it is available. I utilize the concept of "local biologies" to assist in the explication of reports about the embodied experience of physical sensations, including those of well-being, health, and illness. The body should not be assumed to be universal in kind, simply layered over by sociocultural flotsam, as has happened all too often to date.

The concept of local biologies encourages an analytical sensitivity about the situatedness of knowledge and practice and, furthermore, to the contingency of material bodies. Given the remarkable extent of human movement over time, whether voluntary or forced, local biologies travel, resulting in endless transformations and modifications due to ever-changing environmental variables. Not everyone migrates, certain biologies remain local, but in-migration may bring about changes. Given that environments have always already been subjected to modification by means of human activity and with ever-increasing rapidity today, the concomitant biological responses among humans are on the increase. With these admonishments in mind we turn now to the embodiment of epigenetic effects.

Permeable Bodies and Toxic Living

For a decade or more, researchers have been working to elucidate the effects on neurodevelopment of exposure to neurotoxins in utero and early life. Recent work has highlighted epigenetic effects and an apparent intergenerational after-math of such exposures. A 2006 review of an array of 201 neurotoxins, ranging from arsenic to benzene and PCBs, concluded that exposure to hundreds of industrial chemicals are potentially damaging to the developing brains of children worldwide, although it is noted that both timing and the amount of exposure are significant. A 2011 review notes that every year more than 13 million deaths are due to environmental pollutants. Evidence links these pollutants to epigenetic marks associated with a range of disease endpoints. It is emphasized that many of these changes have been shown to be reversible and hence preventive measures are feasible. Lead is the most closely researched toxin to date; it has been shown repeatedly that there is no safe level of exposure during the early years of human development and that exposure results in many epigenetic effects. Decreased brain volume is recognized as lead-related brain atrophy, and is most pronounced in males. Research has also shown negative effects of lead exposure on language function.

Markowitz and Rosner graphically describe the ongoing lead paint scandal in the United States that has steadily unfolded for more than a half-century. Over the years, millions of children have been exposed in their homes to potential lead poisoning, although reliable numbers are not available. It is estimated that today over 500,000 children between one and five years old have lead levels above that which policy makers currently regard as a safe level. Reminiscent of the infamous Tuskegee experiments conducted on African Americans, one hundred children, mostly African American, some less than a year old, living in poor family dwellings where lead paint had been used, have been systematically studied for the effects of lead exposure on their development. A judge who presided over a lawsuit described these young research subjects as "canaries in the coalmine."

It has been shown that lead released from a woman's bones during pregnancy can increase risk for preterm deliveries and low birth weight and, further, affect gene expression in infants involving changes to DNA methylation that may well have lifelong effects. One researcher is quoted as stating: "Lead exposure, rather than a poor social environment, is a key contributor to ... subsequent cognitive and behavior problems." Such a claim prioritizes one variable over another, causing a distortion. It is highly likely that lead exposure does irreparable harm to all humans, but those individuals who are at the greatest risk of being exposed are almost exclusively economically deprived. In 2014, in the impoverished town of Flint, Michigan, with a population of 100,000, nearly 60 percent of which are African American, a water crisis exploded. It became clear that between 6,000 and 12,000 children had been extensively exposed to lead contamination when, in order to save money, the Flint water source was changed from a safe source to one involving use of aging pipes linked to the Flint River that leached lead into the water supply. As one commentator stated: "Some of the darkest chapters in American industrialization are written in lead."

Thanks to the intrepid battle fought by Rachel Carson, DDT was banned in the early 1970s, first in the United States and then worldwide, although it continues to be used in certain malarial regions. And use of PCBs (polychlorinated biphenyls) is banned or severely restricted in most countries, but about 10 percent of the PCBs produced since the late 1920s remain in the environment today.

They are released into the environment primarily from incinerators and build up in the fatty tissues of animals living in water or on land and are passed along the food chain to humans. Dioxins are found throughout the world and accumulate in the food chain, mainly in the fatty tissue of animals. They are highly toxic and cause reproductive and developmental problems, damage the immune system, interfere with hormones, and cause cancer. Dioxin enters the environment primarily from incinerators; of the 419 types of dioxin-related compounds, thirty have significant toxicity.

The effects of PCBs, dioxin, and other toxins in the Arctic are more devastating than elsewhere. Legislation against these chemicals is not effective in the extreme north as yet because toxic residues slowly drift toward the Arctic and accumulate there, making it one of the most contaminated places on earth. The body fat of seals, whales, and walruses hunted for food is highly contaminated, as is the breast milk of many Inuit women. An Inuit grandmother, politically active in circumpolar meetings, is quoted as stating: "When women have to think twice about breast feeding their babies, surely that must be a wake-up call to the world." The situation is exacerbated because the cost of store-bought food is beyond the reach of many Arctic residents.

An article in the *New Yorker* about Tyrone Hayes, a biologist working at the University of California, Berkeley, told a very disturbing story about one of the world's largest agribusinesses, Syngenta. Sales of the herbicide atrazine made by Syngenta are estimated to be around $300 million a year. Hayes's research findings about the toxic effects of atrazine resulted in what appears to be a systematic effort on the part of Syngenta to debunk his findings and destroy his career. Furthermore, the American chemical industry attempted to quash the findings of biologist Frederick vom Saal on bisphenol A, a chemical found in hard plastics and in the coating of food and drink packages, high doses of which have been proven to have a negative impact on health. Moreover, the present partisan government in Canada refuses to legislate effectively against asbestos, although it is the top workplace killer. Clearly, efforts to rid the environment of such chemicals are fraught with dangers other than their toxicity. Rachel Carson encountered great opposition to her work fifty years ago, and it seems that little has changed. [...].

Agent Orange—Lasting Effects in Time and Space

Based on many years of fieldwork that commenced in 2003 in Hanoi, Vietnam, the Danish anthropologist Tine Gammeltoft has documented the devastating effects on reproduction caused by the chemical defoliant Agent Orange that persist more than forty years after the war that lasted from 1962 until 1971. Throughout the war, the US military conducted an aerial defoliation program that was part of a "forced urbanization" strategy designed to force peasants to leave the countryside, where they helped sustain the guerrillas, and move to the cities dominated by US forces. Nearly 20 million gallons of chemical herbicides and defoliants were sprayed onto Vietnam, eastern Laos, and parts of

Cambodia, destroying all plant material in two days. In some areas, toxic concentrations in soil and water became hundreds of times greater than the levels considered safe in the United States.

Agent Orange contains the highly toxic chemical dioxin, known to have long-lasting effects on the environment and human tissue. Gammeltoft documents a widespread fear about the so-called dioxin gene, widely believed by many people living in Vietnam today to be increasing in the population over time. It is estimated that at least 3 million citizens in Vietnam suffer from serious health problems due to exposure to defoliants, and the rate of severe congenital abnormalities in herbicide-exposed people is reckoned at 2.95 percent higher than in unexposed individuals. The media has reported cases of third-generation Agent Orange victims, in which individuals exposed during the war have produced apparently healthy children whose grandchildren are born severely disabled. Animal research has shown that, following fetal exposure, dioxin reprograms epigenetic developmental processes the effects of which may become manifest throughout life and intergenerationally.

Vietnam was given membership in the World Trade Organization in 2007, one result of which was heightened concern by the Vietnamese government about the international visibility of the health of the population as a whole. It was at this juncture that extensive use of ultrasonography was introduced—a political tool designed to ensure the birth of healthy children. One result has been that ultrasound is now used repeatedly during pregnancy as part of antenatal care, even though the Vietnam Ministry of Health does not recommend this practice. Reaching a decision to have an abortion if a deformity is detected by ultrasound is not easy, particularly because many affected families think that abortion is an evil act. Furthermore, everyone involved knows that it can be difficult to assess the extent of the deformity from ultrasound images, although it is equally the case that frequently it is all too evident. Some families, reluctant about abortion, and longing for a healthy child, are raising three or four children with deformities, the most common of which is hydrocephalus ("water on the brain"), which causes severe retardation. A few women discover very late in a pregnancy that their fetus is not normal, and some opt for a late termination, to the great discomfort of their doctors.

Gammeltoft's moving interviews with affected families make clear that many people choose not to entertain the idea that an anomalous fetus detected by ultrasound, or the birth of a horribly deformed child, is due to Agent Orange. They are all too well aware that the stigma attached to Agent Orange families ensures that finding marriage partners for healthy members of the family would be virtually impossible. Better to claim publicly that the anomaly resulted from a common cold that the mother had, or from the heavy work she did while pregnant.

A range of severe illnesses is associated with dioxin exposure, including deadly cancers, Parkinson's disease, and spina bifida, in addition to those associated specifically with pregnancy. Vietnamese researchers have reported these findings, but the official US position is that there is no conclusive evidence that herbicide spraying caused health problems among exposed civilians and their children. However, following extensive lobbying over many years, in 2014 the US Congress passed a five-year aid package of $21 million that amounted to a modest sum for each Vietnam veteran. These cases were settled out of court and no legal liability has ever been admitted. The official position to this day is that the government was in effect prodded into settling these legal suits and that no evidence exists that Agent Orange caused harm. This position is supported by its principal makers, Monsanto and Dow Chemical companies. Children born to Vietnam War veterans who have severe birth defects have received no compensation.

In Vietnam, officials were reluctant to press complaints about Agent Orange because uppermost were concerns about the economy as a whole, notably a desire not to damage the marketing of numerous agricultural and aquacultural products made in Vietnam today. In the mid-1990s, Vietnamese writers and artists finally began to express concern about Agent Orange, and eventually Vietnamese citizens filed a class action suit in the US District Court in New York that was abruptly dismissed. But demands for responsibility are increasingly being heard, spearheaded by nongovernmental organizations. This example illustrates how local biologies travel and are not necessarily stable.

Failing Kidneys and Organ Shortages

In Egypt, when considering receiving an organ from a living related donor, patients have been concerned above all about the sacrifice of the donor and their probable loss of health. In addition, the investment in a transplant affects the economic well-being of the family as a whole and, in addition, the majority of potential recipients believe that their own post-transplant lives would in all probability not be improved enough to justify the involved costs. Egyptian dialysis patients come to such conclusions for several reasons, among which is their awareness that a transplant is unlikely to be a lasting success due to excess pesticide usage in daily life and to improper handling of toxic waste. In Sherine Hamdy's words, Egyptian dialysis patients describe a "local biology" that in effect renders organ transplantation inefficacious due to contaminated environments, endemic parasitic infections, poverty, regular intake of poisonous food, inappropriate use of pharmaceuticals, and medical mistreatment.

The numbers of individuals with kidney failure and liver disease are on the increase in Egypt, and nephrologists note that the kidney diseases they see today are more aggressive than formerly and increasingly difficult to treat. Several of these specialists, like their patients, insist that this situation is due to environmental degradation and the regular ingestion of toxins. Compounding matters, formaldehydes are used to preserve milk, hormone-pumped chickens are imported, rather than using local "country" chickens, and pesticides rejected in the West as unsafe are dumped on Egypt. "Aid" packages from the United States ensure that people eat bread made from imported US wheat that, after storage in the hot and humid conditions of Egypt, produces a fungus associated with kidney and liver toxicity and with cancer. Substantial scientific evidence exists to support these claims, and in addition pesticide residues are present in the local crops.

In classical times Egypt was the breadbasket for the Roman Empire and, even though there were on occasion bad harvests, malnutrition, and corrupt regimes in which grain was not equitably distributed, until recently Egypt has been largely self-sufficient in grain production. The building of the Aswan High Dam, completed in 1971, ruined many hundreds of acres of fertile soil, but even so, the country continued to be able to sustain itself. In recent years the situation has changed dramatically. Since the 1970s, a combination of extensive US subsidies that place Egypt under pressure to import US grain, together with the shift in use of locally grown coarse grains for use as animal food, rather than for human nourishment, has left the majority of Egyptians in difficult straits. Furthermore, the meat products that result from the use of grain to feed animals feed tourists, resident foreigners, and wealthy Egyptians, but not the majority of the population.

Current government expenditure on health care amounts to 2 percent of the total GDP, and the cost of health care for Egyptians is increasing rapidly. Given the expense and justified concerns about toxins, many patients conclude that having a transplant will change their lives little, because the new kidney, too, will soon be damaged. And under the present circumstances donors, for all their benevolence, are unlikely to survive on one kidney, no matter what the medical profession claims. Hamdy concludes that when the patients she talked to reiterated "the body belongs to God" this was not a pious mantra, but rather a resigned grievance about state mismanagement of resources, exploitation of the bodies of the poor, corruption at nearly every turn, and an increasingly toxic environment.

The Egyptian story is by no means unique: in Mexico, the leading cause of death for women today is diabetes that frequently results in kidney failure. In addition, kidney disease is caused by exposure to pesticides, exacerbated by the deregulation of agriculture, waste materials from manufacturing, and pharmaceuticals. Millions of agricultural workers in Asia and Central America also have high rates of kidney failure and mortality due to exposure to agrichemicals. [...].

References

Carson, Rachel. 1962. *Silent Spring*. New York: Mariner Books, Houghton Mifflin Harcourt.
Markowitz, Gerald, and David Rosner. 2013. *Lead Wars: The Politics of Science and the Fate of America's Children*. Berkeley: University of California Press.

DISCUSSION QUESTIONS

1. How does your body affect the ways the medical community may treat you?

2. What role should empathy play in the treatment process?

3. What do medical decision-makers miss by ignoring patient experiences?

4. How does phenomenology relate to bioethics? How could you use phenomenology in understanding the experience of another? How would you use phenomenology in understanding an illness?

5. What environmental factors have impacted health in your community? What about societal factors?

6. How has the history of racial disenfranchisement been shown in medical bias? What about gender? What about sexual preference?

7. What would you want decision-makers (doctors, public health officials, etc.) to know about your experiences as they impact potential decisions they make on your behalf?

Recommended Cases

Henrietta Lacks—Lacks was a Black woman from whom tissue was taken without her consent by doctors at Johns Hopkins, where Lacks was a patient. The tissue was given to a researcher who discovered the cells within would replicate and survive; her cells, named HeLa cells, were said to be immortal. Countless patents were obtained using these cells, an exorbitant amount of money was made off of Lacks's cells, and, while her name was attached to them and her privacy was stripped, she and her potential beneficiaries received nothing from the results of tissues taken from her.

Elizabeth Bouvia—A patient with limited movement due to cerebral palsy, Bouvia wanted to die. Her father, who would not care for Bouvia after she turned 18 due to her disabilities, drove her to a hospital, where she asked to die. Doctors refused to let her die, force-feeding her rather than letting her starve herself and taking her to court to remove decision-making capacity over herself from her.

SARS-COVID-19 Statistics—According to the CDC, you are (as of 09/07/2020) 2.6 times more likely to contract SARS-COVID-19 if you are African American. The contraction rate is 2.8 times higher for Hispanic, Latinx, and Indigenous persons.

Male Birth Control—In 2016, *The Journal of Clinical Endocrinology and Metabolism* published trial results for a male birth control pill. The trials had to stop because of the pill's side effects, which included acne (sometimes severe), mood swings, and depression. These are also some of the many side effects that women who use birth control may endure.

UNIT VI

Euthanasia and Physician-Assisted Suicide

....

Unit VI Introduction

The right to self-determination is inherent when we discuss decision-making about one's life, but what about one's death? This unit focuses on what can often be a final choice. Bringing about one's death has been an ethical topic for millennia; the Hippocratic Oath, for example, forbade euthanasia and physician-assisted suicide. Though these euthanasia and physician-assisted suicide (PAS) seem to be used interchangeably, they are two different processes. Though both describe the voluntary death of an individual, the person who causes the death changes. In euthanasia, the doctor has an active role. In PAS, the doctor prescribes a lethal medication for the patient to use (or not to use) when they choose.

In the United States, some states allow physician-assisted suicide, provided patients meet certain criteria (e.g., expecting death within a certain time frame, multiple physicians involved in decision-making, etc.). What is deemed medical in one state or country is considered murderous in another. David Goodall, a 104-year-old botanist, decided he had lived too long. Euthanasia was illegal in his homeland of Australia, so he flew to Switzerland where he would have the right to his death. Beethoven's "Ninth Symphony" played in the background as he turned a wheel to release the *pharmakon* into his IV.

"Euthanasia" gives an overview of many of the issues that are to come in this unit: balancing individual autonomy and the role of killing, the cost of terminal treatment, and more. Additionally, the reading introduces us to international legal bodies and rulings on euthanasia. While focusing on the legality of these practices, we will also examine the technological developments for extending life and the ethical issues that arise from them.

We focus on the practitioner's role in "Physician, Stay Thy Hand!" While sidestepping the argument regarding an individual's right to death, we examine the history of the medical profession. Its history shows that trust is at its core. Though death may be a right, physicians betray the public trust by involving themselves in death.

"Physician-Assisted Suicide: Safe, Legal, Rare?" describes the shift from the view of PAS as resulting from a lack of pain control toward death becoming a more active process. Should one be able to choose a date the way they would for a wedding, or should we see death as a matter of fate? Moving between international perspectives and social conditioning, we are brought to the question of how to understand death.

READING 6.1

Euthanasia

Amelia Mihaela Diaconescu

T he boundaries of the right to life are difficult to determine. Part of this problem is the existence
or not of an alleged "right to die with dignity." The European states, for example, approach
this problem differently, although they are all part of the European Convention on Human Rights.

Euthanasia is deliberate killing committed under the impulse of compassion in order to relieve
the physical pains of a person suffering from an incurable disease and whose death is, therefore,
inevitable.[1] Euthanasia existed in the human history from ancient times, being tolerated or rejected,
as it happens nowadays.

The difference between euthanasia and assisted suicide lies in how to perform this action.
In the case of euthanasia, the physician administers the lethal medication himself. In the case of
assisted suicide the patient administers himself the medication recommended by the physician.
The medication is specially adapted to precipitate death and to minimize the suffering as much as
possible. This is a characteristic that differentiates the physician assisted suicide from other medical
decisions to suppress life.[2]

Within the current law, euthanasia may be related to the question of homicide incrimination and
especially to the position of the passive subject of the crime. The movement in favor of improving
the condition of the patients in terminal phase, in order to avoid that they endure long suffering and
prejudice their dignity, has manifested along with the movement in favor of abortion. Euthanasia
brings into conflict two groups, called by the Americans *pro life* and *pro choice*, groups outlined
in the United States ever since the 1970s. Therefore, this hypothesis may not contain the cases of
patients whose vital functions are maintained with mechanical assistance. The issue of expression
of a deliberate and prior consent appeared mostly in the case of patients with mental disability.

1 Antoniu, G. C. Bulai, and Gh. Chivulescu (1976), *Criminal Legal Dictionary*. Bucharest: Scientific and Encyclopedic Publishing House, 107.

2 Hecser, Lorand (2001), "Euthanasia—Medical and Socio-Legal Reflections," *Law* no. 11: 93.

In Massachusetts, it was decided that when the patient is unable to express his consent, due to mental disability, the physicians are obligated to bring the case to courts in order for them to decide on the solution allowed.[3]

However, we have to make reference to cases such as Dr. J. Kevorkian's case who was sentenced by the Court of the State of Michigan, on March 26, 1999, for the crime of murder, consisting of carrying out some acts of euthanasia in 130 cases.[4] The attitude towards the dying people constitutes a "carrefour" issue between humanities, religion, law and medicine. In such states, deliberate killing is what was called euthanasia during ancient times, sweet death upon request, nowadays considered death with dignity.

Along with the ageing of the population and progresses in medicine, a significant number of very sick people are maintained alive, thanks to various techniques, which creates an inconstant boundary between life and death. The scientific development of these medical techniques raises complex issues and engenders debates: social debate regarding the high cost of treatment in terminal phases, ethical debate because the struggle for life is a fundamental requirement of our societies and medical ethics in particular; religious debate "to not kill," and legal debate, the right to respect for life, which prohibits and prosecutes homicide; but also the right to respect and guarantee the individual's autonomy, the self-determination of the subject with regard to his own body.

In the case of the euthanasia notion we distinguish three forms: voluntary euthanasia, non-voluntary euthanasia and involuntary euthanasia. Voluntary euthanasia—when death is caused upon the request of the suffering person, non-voluntary euthanasia—when one ends the life of a person who cannot choose by himself between living and dying; involuntary euthanasia—when euthanasia is performed on a person who is able to provide informed consent, but does not, because he was not asked.[5]

Furthermore, we can talk about active euthanasia and passive euthanasia. Active euthanasia is when the death of the patient is caused by a lethal injection and passive euthanasia is when death is due to inaction (stopping the treatment or stopping the technical medical assistance). The first form is not accepted, as it cannot be included in the broader framework of the right to private life. And nevertheless there are movements that demand its legalization, such as the one founded back in 1835 (!) in England and extended later to other countries, affiliated to the World Federation of right to die Societies. As a matter of fact, this form, in some authors' opinion (see Marange), was used by the Nazis, for whom the entire medical practice had become technical. The use of poisonous gas-carbon monoxide and then Zyklor B- was considered to be humane killing. It is shocking the statement of the Nazi neuropsychiatrist Wermacht when asked if he suffered from anxiety or had such dreams—as a

---

3 The Supreme Court in Massachusetts, Saicewicz c. Superintendent of Belchertown State Hospital, 1977.

4 Hecser, Lorand "Euthanasia—Medical and Socio-Legal Reflections," *Law Journal*, 11.

5 For example, causing death by administrating an ever-increasing dose by the doctor who does not intend to cause death, but to alleviate the pain, but which could produce (therefore involuntarily) the patient's death.

result of the slaughters he assisted to or his colleagues' confessions whom he treated—he replied that he had never killed, his quality as a physician allowing him the position of outsider.

The role played by biologists and physicians in drafting the Nazi theories and later in selecting the victims and their extermination represented a sample of the crisis of medical ethics. Euthanasia was practised on tens of thousands of mentally sick people in Germany between the years 1933–1945.

The history of orthonasia begins in 1976, when the Parliamentary Assembly of the Council of Europe analyzed the aspect of the validity of the statements expressing the consent of the persons who gave the doctor the power to give away any measure known in the situation to put an end to the functions of the brain. In the resolution 779/1976 it is demonstrated that the purpose of medical practice is not exclusively intended to prolong life but also to recognize the patient' right not to suffer unnecessarily. In 1977, the Swiss Academy of medical sciences adopted a directive in accordance with which the physician, in the case of persons whose condition is irreversible and who could not be self-conscious, is free to stop using any of the therapeutic resources meant to prolong their life. On the jurisprudential plan, one of the examples is the decision of the Supreme Court of the State of New York that gave the authorization to disconnect the respiratory device in the case of an 83 years old ill person.

Euthanasia, in its passive form (orthonasia) as acceptance of the will of God has been accepted by the Roman Catholic Church which has defined it as inaction or omission which by itself or through its intention causes death.[6] Pope Pius XII claimed that a treatment should not be used unless a person's health will improve to a level at which he can contribute to achieving the highest spiritual values of life or to prevent him in the future to act in this manner.[7]

From the biological and physiological points of view nowadays we accept death as a phase in the movement and transformation of the matter. The question which arises is whether in the cases of patients in terminal phase, we are allowed to accelerate death. The opinions have been and still are polarized around "yes" and "no." Plato recommended abandoning the sick people to death,[8] the Hindus entrusted the elderly suffering patients to the Ganges, and the elderly Eskimoes exposed themselves to the cold when they became dependent on the community. Euthanasia was defended by famous personalities such as: Voltaire, Th. More, Fr. Bacon, D. Hume, B. Shaw, B. Russell, Nobel, and nowadays by a great number of Nobel laureates, but the idea was born along with human consciousness of the suffering and death (the man is the only being with the consciousness of death; Kant qualified him as "metaphysical animal").

---

6 Sacred Congregation for the Doctrine of Faith, Declaration on Euthanasia, Vatican, 1980, 6.

7 Pope Pius XII, (1957), *The Prolongation of Life*, Vatican, 396.

8 Platon *Complete works*, p. 111: "Those whose body is badly built as well as those who have perverted heart and irremediable by nature, will be let to die."

The pro euthanasia arguments are:

- of humanistic nature "right to die alive" and "the right not to suffer unnecessarily": "There is a difference between live and be alive." Being alive is only a biological phenomenon. Animals and even plants are "alive." But living means much more. Living involves both biological and biographical aspects when sufferings darken the joy of a being to live a human life; he is entitled to the same compassion on the part of the society, which it never declines to an animal.[9]
- of medical nature: when medicine is defeated and its role can no longer be the one to save life, it must at least be able to alleviate the suffering.
- of legal nature based on the human being right to dignity and respect for his or her individual autonomy, self-determination of the subject with regard to his own body.
- of economic and social nature: it is about the high costs of healthcare and
- treatment for ill patients in terminal phases, costs that the social debate considers as inadmissible with regard to the discrimination in contrast with the amounts allocated to other categories of persons (e.g., children in centers for orphans).

Worldwide there have been founded numerous civil societies which advocate for euthanasia, the concept of "Not to be resuscitated" was launched and it is permitted in the United States, new rights have arisen within the debates on human rights, rights such as: the right to die with dignity, not to suffer unnecessarily, and others, a rich literature justifying euthanasia has been written.[10] The liberalization of concepts with regard to euthanasia is active in all countries, especially in the well-developed ones, by going down to the legalization in the Netherlands, Austria, Oregon.

There are also numerous points of view against euthanasia; these are in particular of ethical-religious nature, starting from the idea that life is sacred and from the commandment "do not kill," but also of medical nature. Medicine will not ever be able to refute its existence notion: defending life and even, where possible, regaining it. The assertion that euthanasia is only a way of asserting the inability of medicine is fought by those who argue that some incurable diseases may not always be established with certainty from the medical point of view, even assuming that this incurable character would be definite, the practice of euthanasia would give the physician a kind of sovereignty, which would be rather contrary to his real role. Dr. E. London, the president of the World Medical Association (WMA) stated: "It is only God that has the power to decide when life has no value. If we accept the fact that man has the right to decide who of those around him has to die and who has to live, it means that we already find ourselves on the way to concentration camps."

But the achievements of science determined other problems. Medical improved technologies allow to physicians to maintain alive patients who, one or two decades ago would have been dead because these means did not exist at the time. And there comes the question: should physicians always do

---

9 Segal, Frich (1993), *Physicians*. Bucharest: Medical Publishing House, 601.

10 Rachels, J. *End of Life: Euthanasia and Morality*. Mauchester W. *Glory and Dream*, Schwartzenberg, L. *Requiem for Life*, Finkelstein, A. *Euthanasia*, Uffelmen, M. *Euthanasia und Enlosung Schewerkronker*, etc.

what it is possible in order to save their patients' life? It is necessary this heroic effort whose result is the prolongation with several weeks, days or even hours of pain. We ask ourselves if they have to always apply an active treatment to newborn children with extremely serious malformations, even if all they do is adding to their short life more continuous pain. Most of specialists agree that there are some situations when the treatment of maintaining the patient alive must be interrupted and the patient should be left to die in peace. What it was called the torture of terminal therapeutic obstinacies or the type of modern crucifixion (with two oxygen cylinders, with a device for heart support, with a perfusion in the right arm and a transfusion in the left arm or connecting a calf with the artificial kidney, etc.) puts to the issue the limits of human boundaries of science that bioethics, as a bridge between science and theosophy has to solve.

The debate on voluntary euthanasia generally refers to the autonomy borders and to its value in connection with other social values: do we have the right to die if we decide this and the wish to live disappears in order to leave room for the wish to die? The right to life, the right to death, there are two opposite intercrossing notions. Does our right to self-determination, our autonomy to take decisions, allow the agreeing to legal rule of criminal law, whose purpose is to sanction an attitude that is being judged from the social point of view, with the right to decide upon someone's life?

The concept of human rights as a new ethical ideology upon which life is inalienable, sacred and intangible, as well from the physical and spiritual point of view, also determines the legal point of view on euthanasia.

All constitutions consider life as being dignified, not death that is why most of laws incriminate the non-assistance of a person in danger. But, although it was incriminated in most states, euthanasia generally has resulted in merciful decisions, with symbolic sentences or suspensions of sentences, but especially when the action was made because of a merciful feeling and without any material interests. But most legal advisers revealed that the legitimation of euthanasia would generate new racist conceptions, which is reflected in the majority of countries that incriminate euthanasia. Decriminalization of euthanasia took place in Holland, Australia, Oregon (USA), after long debates and points of views that are expressed by public opinion.

Actual directions in the world are in favor of passive euthanasia (renouncement to therapeutic insistences), Resolutions of European Council[11] proclaim the right to die in a dignified manner, relaxed and if possible in comfort, in compliance with the patient's biological will.[12] In vegetative death (brain death) the family can ask the interruption of reanimation. But, at the same time, the Resolutions of European Council ask that the physicians and legal advisers be the first ones to comply with the human rights and note their breach.

---

11 Resolution 613/1976 relating to rights of ill persons and dying persons, Recommendation 779/1976 relating to the rights of ill persons and of dying persons.

12 Biological will—document drawn up during the state of physical and mental state of health of a person through which it supports the refusal to be treated in a certain way, or in any other way, in case of disease.

The authorization of interruption of reanimation is recommended to be done by commissions formed by legal advisers, churchmen, psychologists and physicians, but when defining the moment of death in such cases we must take into account the patient's best interests.

Moral risks towards the social messages of medicine to look after, to form a social unjustified conception on the effects of medicine in deciding on some problems, so serious as life, death and the limits between them, psychological (in terminal phases, given to depression, conscience never being rational) and especially legal (hiding under euthanasia some voluntary crimes, having material purposes, and especially the risk of skipping from legitimation of euthanasia to genocides) oppose in definite and pertinent way to legalization of euthanasia.

In absence of legal texts where it is made reference to euthanasia, legal uncertainty continues to exist, and active voluntary euthanasia places then in an apparent severity of law and a certain legal tolerance. This situation gives in fact room to a real but occult practice that is brought in court. The absence of norms determines a great doubt regarding euthanasia and generates a serious hesitation of the persons that risk remaining in fact without a real protection of their lives and without the right to express their own wish. The society, in lack of some precise rules, can orient its options with no connection to the patients' wish. That is why it seems compulsory to offer to the patient that demands euthanasia legal protection of his freedom to choose and to determine the necessary guarantees in view of keeping his individual freedom, the autonomy in taking decisions regarding his existence and his protection against arbitrary decisions.

In French criminal law there are two doctrinary point of views in what concerns euthanasia—one is considered homicide or ordinary murder; the other makes from the purpose of euthanasia of the crime a legal extenuated excuse, having as a consequence the lowering of the sentence that is stipulated for murder in general.[13]

Italian criminal code (art. 579) and Danish, Icelandic and Finn codes are less severe with voluntary euthanasia. It is the same situation for Austrian criminal code (law from 1934), provided that the patient's request be serious and expressly stated. The same for the Greek code, provided that: the patient's request be serious and insistent, and the author's act be free in his action and inspired by compassion. German criminal code qualifies in a different way the mercy killing. Swiss criminal code accepts to be taken into account the honorable reason.[14] In Romanian legal doctrine it was expressed the opinion that the euthanasia acts could remain not sanctioned if the courts applied a certain reasoning. In the situation of ill people with physical and mental sufferings that cannot be healed and who can no longer be assisted by a physician, it can be argued in favor of no criminal sanction for the physician. In this respect it can be invoked the necessity state regulated by article 45 of Criminal Code. Therefore, there is a permanent social danger for the person's health and integrity.

---

13 Corruti, F. R. (1983), *Etude des aspectes juridiques et medicaux de l'euthanasie et de l'echarnement thérapeutiques: le droit des mourants*, Thèse Université de Bordeaux, 223.

14 Corruti, F. R. *op. cit.*, 225–226.

The action of removing this state, euthanasia is the only means of stopping the suffering, so therefore it is necessary. It is at the same time proportionate if death is compared with the prolongation of suffering for the person and for his family. This reasoning, in the author's opinion must be applied only in the situation where it is certainly determined that the disease is beyond cure and if there are some sufferings that are out of the ordinary, and if it exists the victim's consent.[15]

It is indeed admitted that the right to life does not include a right to die, and life, as supreme social value does not only concern the person, but the state and society, through its social implications, in the sense of legalization of this practice. This dilemma was sometimes solved on legislative way, in the sense of legalizing this practice. It is obvious that in the countries where national courts chose not to hold liable those who practiced euthanasia, they preferred that this practice be regulated as strictly as it may, in order to avoid abuses and to eliminate the risk of some contrary solutions.

Among countries that legalized euthanasia, in contemporary epoch, we mention Holland.[16] Dutch law defines euthanasia as medical active intervention in the purpose of interrupting the patient's life, at his request. In order to frame this hypothesis it is necessary the existence of the patient's consent, in any other case this practice remains a criminal offence.[17] In 1994, in Oregon it was adopted through referendum, following a popular initiative, The Oregon Death with Dignity Act, law that authorizes and stipulates the procedures regarding the physician assisted suicide. Australia followed at short time this example in 1995. In 1997, Columbia was the first country on the American continent that legalized euthanasia.

The right to decide the moment of our own death was related to the right to dignity, notion that does not appear in the text of European Convention on Human Rights, but which was applied in more recent instruments, as for example in the first article of the Charter of Fundamental Rights of the European Union. The respect of human dignity would include the right to live with dignity and to end his own life or to ask for the help of a third person to end his own life. In the conception of those who see euthanasia as a means of not bearing any more the suffering of an incurable disease or the technical means of life prolongation, the problem is debated in article 3 of the Convention which interdicts the sentences or inhuman or degrading treatments.

Legal bodies appointed by European Convention, avoided expressing themselves regarding this right. In case of Widmer vs. Switzerland,[18] they established the fact that the state did not breach its positive obligation to take appropriate measures for life protection, even though it skipped to incriminate passive euthanasia. In a subsequent case, Diane Pretty vs. United Kingdom, which had a great importance to this subject, European Court rejected all the arguments invoked and pronounced in the sense that article 2 could not be interpreted as the recognition of a diametrical opposed right,

---

15 Chiriță, Radu (2001), *Constitutional Right to Life and Criminal Law*. Babeș Bolyai Press, 94.

16 Law regarding Euthanasia and Assisted Suicide, 2001.

17 Chiriță, Radu, *op.cit.*, 123.

18 European Commission on Human Rights, February 10, 1993.

that is to die, without distorting the terminology used. In this case, as we have already shown, it was about a 43 years old woman, who suffered from an incurable disease that completely paralyzed her and she was in agony.

The jurisprudence of ECHR sustains that, even if the right to die is not a component of the right to life, at least not in the meaning of article 2 from the Convention, it does not restrain the states to decide the sanctionary regime that they will apply to these acts. In other words, it is about a domain that still remains in the edge of appreciation of national lawmaker. Referring to interdiction of euthanasia there are some contradictions in legal practice of states, and of international bodies in respect with human rights. This contradiction comes from the conflict between the freedom of conscience and the right to life, where the freedom of conscience prevails. Therefore, a person can refuse an absolute necessary treatment for saving life, without forcing the person to follow it, motivating the existence of some religious beliefs for example, even if this endangers his existence.[19] But when it is about the free and conscious choice to end his or her life, because of some exceptional conditions, this demarche is forbidden. It is true that in this situation we are facing a conflict between a right and a freedom that doctrine and jurisprudence had not unanimously known yet in case of euthanasia. However, even if the most often the doctrine and practice had make any connection between the alleged "right to die" and freedom of conscience, this comparison is worthy to be taken into consideration, in order to determine the limits that the society imposes to the freedom of choice for a person.

Final conclusion: Progresses of science reached a situation of confrontation with ethics; biology influenced man's history and his identity, "but the biological phenomenon must be always depassed in order to form the symbolic part ... today the right is directed in order to rethink the powers the man has on his own body, such as the protection of the body against outside threats."[20]

As well, we have to rethink science and freedom and the limits of science on the rules of conduct destined to people.

Protection of freedom and man's dignity involves an action in main domains of life in society. The right, which is mindful to these new realities, tries to impose changing strategies, citizen's protection of dignity and freedom. In appreciation of social achievements, the legal adviser states as well the need of interdisciplinary knowledge, in order to discover the best means, legal or not, in view of fulfillment of these social purposes.

The right through a precise constitutional norm, through the constitutional exactness as an easy and accessible means of jurisdictional protection of fundamental rights, is the instrument of these changes.

---

19 Supreme Court of Spain, decision from June 27, 1997.

20 Terre, Fr. (1987), *L'enfant de l'esclave. Génétiques et droit.* Paris : Flammarion, 116.

Physician, Stay Thy Hand!

Bernard Baumrin

....

S uicide may or may not be a natural right, that depends on whether there are natural rights and whether one owns one's own being to dispose of as the owner sees fit. Suicide is only a legal right where it has been established by law, customary or explicit. States, in general, prohibit suicide rather than facilitate it. Further, aiding suicide is so fraught with the possibility of serving merely to disguise murder that there is a tendency to legislate against it even more vigorously than against suicide itself.

Enlisting someone to aid in one's suicide endangers the confederate. However clear the one dying might make to posterity that the person aiding was dragooned into the role, suspicions might still linger and, where aiding in a suicide is illegal, the costs might be very great. Can we readily distinguish Pindarus stabbing Cassius at Cassius's insistence from Strato holding the sword that Brutus runs upon? Do we really know that Pindarus and Strato aided in the suicides of Cassius and Brutus rather than that they killed them, perhaps to be free, perhaps to run from the battlefield, perhaps to escape Anthony and Augustus's slaughtering troops? Where are Cassius and Brutus to testify that Pindarus and Strato aided in their deaths at their commands or entreaties? Where is the testimony that they are now to be free? Who is left in danger when the suicide is successful?

That Cassius and Brutus did not wish to face the fate that defeat destined them to (dragged to their place of execution in public humiliation) is quite understandable (neither did Hitler, nor Goering) and that may justify their belief that suicide was a more desirable way to die. And so it may have always seemed—if one must die and, if the time between knowing that and it occurring promises to be one of maximal psychic or physical discomfort, then suicide seems a quite reasonable alternative. Probably the best way to go is quickly and painlessly, but people have chosen to drown, to starve, and to eviscerate themselves. Those who have chosen painless exits, at least relative to the obviously painful ones, have tended to favor poison—it doesn't maim, it can be taken in comfort, and one can preserve one's vision of oneself intact, just merely dead. So, if you want to die by your own hand, peacefully, painlessly, and relatively quickly, you go to an expert in poisons. One versed in the properties of poisons should be able to provide the perfect solution. Such a person, one with knowledge of the various properties of poisons, will be able to provide a quick-acting, hopefully painless potion that one can take at one's leisure.

In the Greek world such persons, ones with knowledge of the properties of potions, poisonous and otherwise, included physicians. And one may imagine that they were called upon often to provide such potions both for murder and suicide. So, with good reason, as an article of their craft, as part of their undertaking among themselves, they forswore providing poisons to patients, presumably regardless of the entreaties they might be subjected to. The Hippocratic Oath included that solemn undertaking, and it implied that physicians were not in the business of ending lives, whether of fetuses or of the distraught.

When Socrates took hemlock, he drank it himself, and it was provided by the prison guard. We might take this case as the paradigm of the good suicide—a drug which causes no pain, which causes a creeping numbness, which permits one to be in control of ones faculties until the last moment, and which one takes oneself. All that remains to be settled is procuring the poison. Effective poisons are readily available today without it being required that physicians prepare them. Moreover, the effects (and side effects) of various drugs and their lethal doses are well known and, in any case, can be readily found in the *Physician's Desk Reference*. The physician as pharmacist is no longer required and need have no role to play in aiding a suicide. If the state wished to legitimize suicide and take a favorable attitude toward drug-facilitated death, it could set up a special pharmacy to dispense suicide drugs (upon proper proof that it was intended for self-use, and of identification). If that idea sounds strange, it is at least partly because we immediately see avenues for abuse, and partly because we can foresee the possibility that lethal potions would lie about unused for some time.

But, in principle, there is no pressing need for physician assistance in the normal case of suicide, and there is a long standing tradition amongst physicians that they do not aid in (and surely do not abet) the suicide of others. I do not extend this tradition to other healthcare workers; first, because they do not have long traditions (except perhaps for nuns who have maintained hospitals for at least several hundred years) and, second, there is as yet no suggestion that persons have a right to their aid in dying (though surely many a nurse has engaged in both involuntary as well as voluntary euthanasia, and one can be sure that many would readily volunteer).

My argument is only about physicians and their duties. Anyone claiming a right to physician assistance in dying must show that some physician has a duty to satisfy that right. Were a legal right to physician assistance in dying created, the physician might still be duty bound to reject the legal claim on several grounds, the most salient of which is his professional obligation not to aid in suicide. This might be founded on an oath, or simply on the assumption that the doctoring tradition he or she has joined does not do that sort of thing. The scientific tradition assumes truth in reporting, and there is justifiable public outrage when there is a breach. The teaching tradition assumes that grades and examinations will be dealt with as objectively as possible, and there is justifiable outrage at trading grades for favors, or passing exams that are clearly failures. The doctoring tradition assumes that the physician will do his or her utmost to heal the sick and to preserve and extend life. That is the "reason to be" for physicians, just as truth is the reason to be for scientists. The doctor trades on an aura developed in antiquity and extended through the centuries by a continuous line of physicians, summed up, if crudely, as "Do no harm." Among the obvious implications of this commanding injunction is that the physician will not do anything that unnecessarily endangers the life of the patient. Only medical necessity, and the paucity of knowledge, permit the physician to endanger the patient's life. There are of course (and there always have been) numerous occasions when the physician does endanger the patient, but never knowingly or intentionally, unless some treatment promises to have a good outcome.

Every physician, I contend, trades on this aura that the present physician did not create. It is the product of his or her predecessors. No one alive today has a right to abjure that aura without publicly and prominently labeling himself or herself a "physician" of a special sort, a doctor for hire who has no prior commitments to the principles of the physician's profession.

The reasoning behind the Hippocratic Oath is to us now merely guesswork, but the fact that it served well for so long suggests that it created a public perception and a public trust that benefited every adherent. From the itinerant Greek physicians in the Roman Empire to Doctors Without Borders, the public trust that physicians are partisans for life and health has opened doors, provided security in times of war, has created a perception of neutrality greater than the neutrality of the cloth. We can see in our mind's eye clerics—priests, ministers, rabbis, imams, etc.—leading their followers into holy battle, but we cannot see that as part of the calling of physicians. The physicians' holy icon is life, not victory; the enemy is death and disease, not people and their states. This allegiance is not any physician's personal property to dispose of. The lesson of the Nazi doctors during World War II, and the outrage that greeted the disclosure of their behavior, is that physicians cannot engage in activities that deliberately endanger the health and life of patients. It is absolutely forbidden, not merely undesirable; the prohibition attaches to them irrevocably in a way that cannot be abjured.

It is not merely qua physician that one must not deliberately endanger others under the guise of doing medicine, but that a physician is to be always a champion of life and health in every context. That is the public perception, and the public perception binds the practitioner as tightly as an oath, as tightly as any obligation can bind.

The contemporary debate about physician-assisted suicide turns on two special kinds of cases that are cited as exceptions to the general principles set out above. I think that those principles are beyond reasonable dispute, but if there is an argument to be made that physicians don't have the duties cited above, then the rest of my argument will be irrelevant because it depends on those principles for its premises.

The two exceptions are the irreversibly comatose and the terminally ill. The cases are actually very different because in the latter there is a range of communication issues entirely absent from the former. In the former case, whether or not there are advance directives about the cessation of care or not commencing care, the cases that we have to deal with primarily are those where the patient is under care and can no longer communicate his or her present wishes about anything. Thus the first exception, the irreversibly comatose, can only metaphorically be considered physician-assisted suicide; rather, it is physician (or someone else) killing. The issue of physician-assisted suicide with the comatose (irreversible or not!) only arises when a patient has left such an explicit advance directive that called upon someone (physician or otherwise) to kill the patient when certain explicitly identified events occurred. Even in this restricted class of case, and one might imagine that there were fairly few, it is a stretch to consider killing that patient a case of suicide. If the patient's coma was possibly reversible then it is unlikely that the advance directive, however explicit, would be taken at its word.

The case of the terminally ill is not straight-forward because the range of complications and the elasticity of the relevant time frame make determining who is to count as terminally ill either too easy or too difficult—too easy because anyone who has an illness or condition which will result in death if not cured is terminally ill, and nevertheless there may be much life left, and life well worth living for anyone who is terminally ill. "Terminally ill" is only superficially a medical term; at base it is a catchall for such states as declining rapidly, irreversibly ill (i.e., incurable with a prognosis of imminent death—but then imminent has a wide range itself), prone to sudden heart attact, stroke, seizure, etc. For many in such condition there is realistic hope that the state they are in can be reversed or at least stabilized, and the threatened decline abated. Of course, for a great many that is not and will not be the case. We do all die, and our deaths are frequently preceded by periods of irreversible decline.

We can then isolate a class of noncontroversial cases: those who are facing the imminence of death, with conditions reliably predicted to be irreversible. It is irrelevant whatever they are or are not in pain; what matters is that they do not want to face whatever the future holds for them, brief as it may be. This desire may be entirely rational, though of course it is sometimes quite irrational. We can assume that there are cases that reasonable people would agree merited serious consideration for suicide. The question then, and the only question that I am addressing, is whether any physician should be involved in implementing that decision.

On the positive side, there are all the obvious issues of patient self-determination (patient autonomy), easing suffering (beneficence), reducing the dangers of unauthorized individuals volunteering, and largely eliminating the chance that things will go wrong and the patient will survive though more maimed than before.

On the negative side, the principal consideration is that aiding a suicide compromises the physician's professional persona. Trust is betrayed, and it is likely that it will be replaced by an image of the physician as "angel of death" (which was what Doctor Mengele was called in the camps). Should physician-assisted suicide be decriminalized, patients will routinely ask for such help and only the mean-spirited will decline. The physician who wishes to perpetuate the tradition of the doctor as the unflagging opponent of death will cease to be perceived as heroic but now as curmudgeonly.

Those who see themselves as duty-bound not to aid in suicide will be castigated as old-fashioned and uncooperative.

Yet, nothing, absolutely nothing, requires that physicians be the instruments of suicide aid. Assuming the would-be suicide cannot muster the courage or get the appropriate drugs, some public-spirited nonprofit group, like the Society for the Right to Die or the Hemlock Society, can be called in to do the appropriate paperwork; obtain the relevant information from the physician or hospital; procure the relevant official government approvals; get the drugs; administer the drugs; and arrange to have the whole thing on videotape. Except for signing the death certificate, no physician needs to be involved and none should. The physician's task is to tell the patient (and relevant others when the patient is either a minor, under guardianship, or comatose) what's wrong, and to the best of the doctor's ability, what is going to happen. The physician's job is to heal the sick, to stave off death, and to say as best as he or she can what the future will be like for each particular patient. The physician gets to be the helpless person's medical guide because he or she is trusted to hold the patient's good uppermost, and the patient's good does not include death.

I have not strayed far from the beginning. Doctors must not engage in assisting suicide. They are inheritors of a valuable tradition that inspires public trust. None should be even partly responsible for the erosion of that trust. Nothing that is remotely beneficial to some particular patient in extremis is worth the damage that will be created by the perception that physicians sometimes aid and even abet people in taking their own lives.

Who knows if, after the legalization of physician-assisted suicide is publicly approved, doctors will also be entreated (by the state, or relatives, or HMOs), to end "lives that are not worth living."

Physician-Assisted Suicide

Safe, Legal, Rare?

Margaret P. Battin

The Way It Looks Now

Observe the current debate over physician-assisted suicide: On the one side, supporters of legalization appeal to the principle of autonomy, or self-determination, to insist that terminally ill patients have the right to extricate themselves from pain and suffering and to control as much as possible the ends of their lives. On the other, opponents resolutely insist on various religious, principled, or slippery-slope grounds that physician-assisted suicide cannot be allowed, whether because it is sacrilegious, immoral, or poses risks of abuse. As vociferous and politicized as these two sides of the debate have become, however, proponents and opponents (tacitly) agree on a core issue: that the patient may choose to avoid suffering and pain. They disagree, it seems, largely about the means the patient and his or her physician may use to do so.

They also disagree about the actualities of pain control. Proponents of legalization insist that currently available forms of pain and symptom control are grossly inadequate and unsatisfactory. Citing such data as the SUPPORT study, they point to high rates of reported pain among terminally ill patients, inadequately developed pain-control therapies, physicians' lack of training in pain-control techniques, and obstacles and limitations to delivery of pain-control treatment, including restrictions on narcotic and other drugs.[1] Pain and the suffering associated with other symptoms just aren't adequately controlled, proponents of legalization insist, so the patient is surely entitled to avoid them—if he or she so chooses—by turning to earlier, humanely assisted dying.

---

1 According to the SUPPORT study, about 50 percent of dying hospitalized patients were reported to have experienced moderate to severe pain at least 50 percent of the time in their last three days of life. The Support Principal Investigators, "A Controlled Trial to Improve Care for Seriously Ill Hospitalized Patients," *Journal of the American Medical Association 274* (20) (1995): 1951–98.

Opponents of legalization, on the other hand, insist that these claims are uninformed. Effective methods of pain control include timely withholding and withdrawal of treatment, sufficient use of morphine or other drugs for pain (even at the risk of foreseen, through unintended, shortening of life), and the discontinuation of artificial nutrition and hydration. When all other measures to control pain and suffering fail, there is always the possibility of terminal sedation: the induction of coma with concomitant withholding of nutrition and hydration, which, though it results in death, is not to be seen as killing.

Proponents laugh at this claim. Terminal sedation, they retort, like the overuse of morphine, is functionally equivalent to *causing* death.

Despite these continuing disagreements about the effectiveness, availability, and significance of current pain control, both proponents and opponents in the debate appear to agree that *if* adequate pain control were available, there would be far less call for physician-assisted suicide. This claim is both predictive and normative. *If* adequate pain control were available, both sides argue, then physician-assisted suicide would be and should be quite infrequent—a "last resort," as Timothy Quill puts it, to be used only in exceptionally difficult cases when pain control really does fail. Borrowing an expression used by President Clinton to describe his view of abortion, proponents insist that physician-assisted suicide should be "safe, legal, and rare." Opponents do not believe that it should be legal, but they also think that if it cannot be suppressed altogether or if a few very difficult cases remain, it should be very, very rare. The only real disagreement between opponents and proponents concerns those cases in which adequate pain control cannot be achieved.

What accounts for the opposing sides' underlying agreement that physician-assisted suicide should be rare is, I think, an unexamined assumption they share. This assumption is the view that the call for physician-assisted suicide is what might be called a *phenomenon of discrepant development*: a symptom of the disparity in development between two distinct capacities of modern medicine, the capacity to extend or prolong life and the capacity to control pain. Research, development, and delivery of technologies for the prolongation of life have raced far ahead; those for control of pain lagged far behind. It is this situation of discrepant development that has triggered the current concern with physician-assisted suicide and the volatile public debate over whether to legalize it or not.

The opposing sides both also hold in common the view that what would lead to the resolution of the problem is whatever set of mechanisms would tend to equalize the degree of development of medicine's capacities to prolong life and to control pain. To achieve this equalization, two simultaneous strategies are recommended: cutting back on overzealous prolongation of life (as Dan Callahan, for example, has long recommended), and at the same time (as Hospice and others have been insisting) accelerating the development of technologies, modes of delivery, and physician training for more effective methods of pain control.[2] As life prolongation is held back a bit, pain control can catch up,

2 See, e.g., Daniel Callahan, *Setting Limits: Medical Goals in an Aging Society* (New York: Simon and Schuster, 1987), *What Kind of Life? The Limits of Medical Progress* (New York: Simon and Schuster, 1990), and *The Troubled Dream of Life: Living with Mortality* (New York: Simon and Schuster, 1993). On acceleration of pain-control development, see especially the work of Kathleen Foley, "Pain, physician-assisted suicide, and euthanasia," *Pain Forum* 4 (3) (1995) and other works.

and the current situation of discrepant development between the two can be alleviated. Thus calls for physician-assisted suicide can be expected to become rarer and rarer, and as medicine s capacities for pain control are finally equalized with its capacities for life prolongation, finally virtually to disappear. Almost no one imagines that there will not still be a few difficult situations in which life is prolonged beyond the point at which pain can be effectively controlled, but these will be increasingly infrequent, it is assumed, and in general, as the disparity between our capacities for life prolongation and for pain control shrinks, interest in and need for physician-assisted suicide will decrease and all but disappear.

Fortunately, this view continues, the public debate over physician-assisted suicide now so intense will not have been a waste, since it has both warned against the potential cruelty of overzealous prolongation of life and at the same time stimulated greater attention to imperatives of pain control. The current debate serves as social pressure for bringing equalization of the disparity about. Yet as useful as this debate is, this view holds, it will soon subside and disappear; we're just currently caught in a turbulent—but fleeting—little maelstrom.

The Longer View

That's how things look now. But I think we can also see our current concern with physician-assisted suicide in a longer-term, historically informed view. Consider just three of the many profound changes that affect matters of how we die. First, there has been a shift, beginning in the middle of the last century, in the ways in which human beings characteristically die. Termed the "epidemiological transition," this change involves a shift away from death due to parasitic and infectious disease (ubiquitous among humans in all parts of the globe prior to about 1850) to death in later life of degenerative disease, especially cancer and heart disease, which together account for almost two-thirds of deaths in the developed countries.[3] This means dramatically extended lifespans and also deaths from diseases with characteristically extended downhill terminal courses. Second, there have been changes in religious attitudes about death: People are less likely to see death as divine punishment for sin, or to see suffering as a prerequisite for the afterlife, or to see suicide as a highly stigmatized and serious sin rather than the product of mental illness or depression. Third among the major shifts in cultural attitudes that affect the way we die is the increasing emphasis on the notion of individual rights of self-determination, reinforced in the latter part of this century by the civil rights movements attention to individuals in vulnerable groups. This shift has affected self-perceptions and attitudes towards the terminally ill, and patients, including dying patients, are now recognized to have a wide array of rights previously eclipsed by the paternalistic practices of medicine.

---

3 The term originates with A. R. Omran, "The Epidemiologic Transition: A Theory of the Epidemiology of Population Change," *Milbank Memorial Fund Quarterly* 49 (4): 509–38 (1971); and the theory is augmented in A. Jay Olshansky and A. Brian Ault, "The Fourth Stage of the Epidemiologic Transition: The Age of Delayed Degenerative Disease," in Timothy M. Smeeding et al., eds., *Should Medical Care Be Rationed By Age?* (Totowa, NJ: Rowman and Littlefield, 1987, pp. 11–43.

These three transitions, along with many other concomitant cultural changes, invite us to see our current concern with physician-assisted suicide in a quite different way, not just as a phenomenon resulting from the currently disparate development of life-prolonging and pain-controlling technologies, a temporary anomaly, but as a precursor, an early symptom of a much more substantial sea change in attitudes about death. We might call this shift in attitudes a shift towards "directed dying" or "self-directed dying," in which the individual who is dying plays a far more prominent, directive role than in earlier eras in determining how and when his or her death shall occur. In this changed view, dying is no longer *something that happens to you* but *something you do.*

To be sure, this shift—if it is one—can be seen as already well under way. Taking its legally visible start with the California Natural Death Act of 1976, terminally ill patients have already gained dramatically enlarged rights of self-determination in matters of guiding and controlling their own deaths, including rights to refuse treatment, discontinue treatment, stipulate treatment to be withheld at a later date, designate decisionmakers, and to negotiate with their physicians, or have their surrogates do so, such matters as DNR orders, withholding and withdrawal of ventilators, surgical procedures, nutrition and hydration, the use of opioids, and even terminal sedation. Some patients also negotiate, or attempt to negotiate, physician-assisted suicide or physician-performed euthanasia with their physicians. In all of this, we already see the patient playing a far more prominent role in determining the course of his or her dying process and its character and timing, and far more willingness on the part of physicians, family members, the law, and other parties to respect the patient's preferences and choices in these matters.

But this may be just the tip of a looming iceberg. For we may ask whether, much as we human beings have made dramatic gains in control over our own reproduction (particularly rapidly in very recent times—the birth control pill was introduced just thirty years ago), we human beings are beginning to make dramatic gains in control over our own dying, particularly rapidly in the last several decades. We cannot keep from dying altogether, of course. But by using directly caused death, as in physician-assisted suicide, it is possible to control many of dying's features: its timing in the downhill course of a terminal disease, its place, the exact agents which cause it, its observers, and so on. Indeed, as Robert Kastenbaum has argued, because it makes it possible to control the time, place, manner and people present at one's death, assisted suicide will become the *preferred* manner of dying.[4]

But this conjecture doesn't yet show what could actually motivate such substantial social change, away from a culture which sees dying primarily as *something that happens to you,* to a culture which sees it as *something you do*—a deliberate, planned activity, one's final and culminative activity. What might do this, I think, is a conceptual change, or, more exactly, a shift in decisional perspective in choice-making about pain, suffering, and other elements of dying. It is the kind of shift in decisional perspective that evolves on a society-wide scale as a populace gains understanding of and control

4 Robert Kastenbaum, "Suicide as the Preferred Way of Death," in Edwin S. Shneidman, ed. *Suicidology: Contemporary Developments* (New York: Grune and Stratton, 1976), pp. 425–41.

over a matter, a shift in choice-making perspective from a stance we might describe as immediately involved or "enmeshed," to one that is distanced and reflective. (I'll use two Latin names for these stances later.) This shift can occur for many features of human experience—it has already largely occurred in the developed world with respect to reproduction—but it has not yet occurred with respect to death and dying. It has not yet occurred—or rather, perhaps, it has just begun.

Take a patient, an average man. This particular man is so average that he just happens to have contracted that disease which is the usual diagnosis (as we know from the Netherlands) in cases of physician-assisted suicide—cancer—and he is also so average that this disease will kill him at just the average life-expectancy for males in the United States, 72.8 years.[5] Furthermore, he is also so average that if he does turn to physician-assisted suicide, he will choose to forgo just about the same amount of life that, on average, Dutch patients receiving euthanasia or physician-assisted suicide do, less than 3.3 weeks.[6] He has been considering physician-asssisted suicide since his illness was first diagnosed (since he is an average man, this was about 29.6 months ago), but now, as his condition deteriorates, he thinks more seriously about it. His motivation includes both preemptive elements, the desire to avoid some of the very worst things that terminal cancer might bring him, and reactive elements, the desire to relieve some of the symptoms and other suffering that he is already experiencing. *It's bad enough now,* he tells his doctor, *and it will probably get worse.* He asks his doctor for the pills. He is perfectly aware of what he may miss—a number of weeks of continued life, the possibility of an unexpected cure, the chance, even if it is a longshot, of spontaneous regression or remission, and—not to be overlooked—the possibility that that the worst is over, so to speak, and that the remainder of his downhill course in terminal cancer won't be so bad. He is also well aware that even a bad agonal phase may nevertheless include moments of great intimacy and importance with his family or friends. But he makes what he sees as a rational choice, seeking to balance the risks and possible benefits of easy death now, versus a little more continuing life with a greater possibility of a hard death. He is making his choice *in medias res,* in the middle of things, as the physical, social, and emotional realities of terminal illness engulf him. He is enmeshed in his situation, caught in it, trapped between what seem like two bad alternatives—suffering, or suicide.

But, of course, he might have done his deciding about how his life shall end and whether to elect physician-assisted suicide in preference to the final stages of terminal illness from a quite different, more distanced perspective, a secular version of the view *sub specie aeternitatis.* This is not just

5 Data on physician-assisted suicide and euthanasia in the Netherlands is provided by what is called the "Remmelink Commission Report." Paul J. van der Maas, Johannes J. M. van Delden, and Loes Pijnenborg, "Euthanasia and Other Medical Decisions Concerning the End of Life," published in full in English as a special issue of *Health Policy* 22, nos. 1 and 2 (1992); and, with Caspar W. M. Looman, in summary in *The Lancet* 338 (Sept. 14, 1991): 669–74; and the five-year update in Paul J. van der Maas, et al., "Euthanasia, Physician-Assisted Suicide, and Other Medical Practices involving the End of Life in the Netherlands, 1990–1995," *New England Journal of Medicine* 335 (22): 1699–1705 (1996).

6 See Ezekiel J. Emanuel and Margaret P. Battin, "The Economics of Euthanasia: What are the Potential Cost Savings from Legalizing Physician-Assisted Suicide?" MSS in progress, citing data from the Netherlands.

an objective, depersonalized view—anybody's view—but his own, distinctively personal view not confined to a specific timepoint.[7] Rather than assessing his prospects from the point of view he has at the time at which he would continue or discontinue his life—that point late in the course of his illness when things have already become "bad enough" and are likely to get worse—he might have done his deciding, albeit rather more hypothetically, from the perspective of a more generalized view of his life.

From this alternative perspective, what he would have seen is the overall shape of his life, and it is with respect to this that he would have made his choices about how it shall end. Of course, he could not know in advance whether he will contract cancer, or succumb to heart disease, or be hit by a bus—though he does know that he will die sometime or other. Consequently, his choices are necessarily conditional in form: "*If* I get cancer, I'll refuse agressive treatment and use hospice care"; "*If* I get AIDS, I'll ask for physician-assisted suicide"; "If I get Alzheimer's, I'll commit suicide on my own, since no physician besides Jack Kevorkian would help me," and so on. Although conditional in form and predicated on circumstances that may not occur, these may be real choices nonetheless, and, particularly because they are reiterated and repeated over the course of a lifetime, have real motive force.

The difference, then, between these two views is substantial. In the first, our average man with an average terminal cancer, doing his deciding *in medias res,* is deciding whether or not to take the pills his physician has given him now. It is his last possible couple of weeks or a month (on average, 3.3 weeks) that he is deciding about. Even if continuing life threatens pain and other suffering, it is still all he has left, and while it may be difficult to live this life—all he has left—it may also be very difficult to relinquish it.

In contrast, if our average man were doing his deciding *sub specie aeternitatis,* from a distanced though still personal viewpoint not tied to a specific moment in his life, he would have been deciding all along between two different conceptions of his own demise, between two possible lives for himself. One of his possible lives would, on average, be 72.8 years long, the average lifespan for a male in the United States, with the possibility of substantial suffering at the end—on average, as the SUPPORT study finds, a 50 percent chance of moderate to severe pain at least 50 percent of the time during the last three days before his death. The other of his possible lives would be about 72.7 years long, foreshortened on average 3.3 weeks by physician-assisted suicide, but with a markedly reduced possibility of substantial suffering at the end. (This shortening of the lifespan is not age-based but time-to-death based, planned for, on average, 3.3 weeks before an unassisted death would have occurred; it occurs in this example at age 72.7 just because our man is so average.) This latter, shortened life also offers our man the opportunity to control the timing, the place, the manner, and other features of his death in the way he likes. Viewed *sub specie aeternitatis,* at any or many earlier points in one's life or

7 The distinction I am drawing here between personal views *in medias res* and *sub specie aeternitatis* is thus not quite the same as that drawn by Thomas Nagel between subjective and objective views, though it has much in common with Nagel's distinction in contexts concerning death. See Nagel, *The View from Nowhere* (New York: Oxford University Press, 1986), especially chapter XI, section 3, on death.

from a vantage point standing outside life, so to speak, the difference between 72.8 and 72.7 seems negligible: These are both lives of average length not interrupted by grossly premature death. Why not choose the one in which the risk of agonal pain—as high as 50/50, according to the SUPPORT study—is far, far less great, and the possibility of conscious, culminative experience, surrounded by family members, trusted friends, and permitting final prayers and goodbyes, is far, far greater?

It may seem difficult to distinguish these two choices in practice. This is because we typically make our decisions about death and dying *in medias res,* not *sub specie aeternitatis,* and our medical practices, our bioethics discussions, and our background culture strongly encourage this. The call for assisted dying, like other patient pleas, is seen as a reaction to the circumstances of dying, not a settled, longer term, preemptive preference.[8] True, some independently minded individuals consider these issues in a kind of background, hypothetical way throughout their lives, but this is certainly not the practical norm. We can only really understand this view as involving a substantial cultural shift from our current perspective.

But if this shift occurs, a slightly abbreviated lifespan in which there is dramatically reduced risk of pain and suffering will not only seem to be preferable to one which is negligibly longer but carries substantial risk of pain and suffering in its agonal phase, it will also be seen as rational and normal to plan for this abbreviated lifespan and to plan the means of bringing it about. The way to ensure it, of course, is to plan for direct termination of life. After all, one cannot count on being able to discontinue some life-prolonging treatment or other—refusing antibiotics, disconnecting a respirator—to hasten death and thus avoid what might be the worst weeks at the end. This most likely means planning for physician-assisted suicide. From this distanced perspective, a 72.7-year life with a virtually assured good end looks much, much better than a 72.8-year life that has an even chance of coming to a bad end. Arguably, it would be rational for any individual, except those for whom religious commitments or other scruples rule out suicide altogether, to plan to ensure this. But if it looks this way to one individual, it will look this way to many; and it is thus plausible to imagine that physician-assisted suicide would not be rare but rather a choice viewed as rational and preemptively prudent by many or most members of the culture. Thus it can come to be seen as a normal course of action, not a rarity or a "last resort." To be sure, there are other ways of abbreviating a lifespan to avoid terminal suffering—withdrawing or withholding treatment, overusing of morphine or other pain-relieving drugs, discontinuing artificial nutrition and hydration, and terminal sedation—but these cannot be used unless the patient's condition has already worsened and is thus likely to involve that pain or suffering the person might choose to avoid. Thus these other modalities function primarily reactively; it is assisted suicide that can function preemptively.

But, as soon as planning for a normal, slight abbreviation in the lifespan by means of assisted suicide becomes conceptually possible not just for our average man but for actual persons in general, it also becomes possible to imagine a wide range of context-specific cultural practices which might

---

8 See C. G. Prado, *The Last Choice: Preemptive Suicide in Advanced Age* (Westport, Conn.: Greenwood Press, 1990).

emerge surrounding physician-assisted suicide. After all, that a person understands and expects his lifespan to be one which will end in an assisted death a few weeks before he might otherwise have died, while he is still conscious, alert, and capable of deciding what location he wants it to take place in, what family members, caregivers, clergy, or others he wants to have present, what ceremonies, religious or symbolic, he wants conducted, etc., suggests that more general social practices would grow up around these possibilities. After all, our average man sees his life this way; but it is possible for him to do this partly because the others in his society see their lives this way as well. Attitudes about death are heavily socially conditioned, and so are the perspectives from which choice-making about death is seen.

This is the precondition for the development of a whole range of social practices supporting such choices. These might include various kinds of practical supports, such as legal, insurance, and other policies that treat assisted dying as acceptable and normal; various sorts of cultural and religious practices that similarly treat assisted dying as acceptable and normal (for instance by developing rituals and rites concerning the forthcoming death); familial supports within the family, including family gatherings, preparing for the death, and sharing reminiscences and goodbyes; pre-death dispositions of wills and life insurance (we already recognize viaticums, pre-death payoffs of life insurance for terminally ill patients); and even such now-inconceivable practices as pre-death funerals, understood as ceremonies of leavetaking and farewell, expressions of both celebration of a life complete and grief at its loss. In turn, such social practices come to function as positive reasons for choosing a somewhat earlier, elective death—formerly and rudely called "physician-assisted suicide," even when pain control is no longer the issue at all—and the new social pattern, so different from our current one—reinforces itself. This has nothing to do with a *Soylent Green* sort of view, in which people are forced into choices they do not genuinely make (this film can be understood only from our current, *in medias res* view); but a world in which their normal choices have genuinely changed, and changed for reasons which seem to them good.

Furthermore, if the culture-wide view of choice-making about death and dying were more fully held *sub specie aeternitatis* in this distanced, less enmeshed and less merely reactive way, in which earlier, elective death becomes the norm, we could also expect the more frequent practice of "setting a date," as people who have contracted predictably terminal illnesses carry out the plans they had been developing all along for their own demises. Setting a date for one's own death—presumably, a couple of weeks or so before the date it might naturally have been, revisable of course in the light of any changes in the diagnosis or prognosis—would still be both preemptive and reactive in character, but far more preemptive than choices made *in medias res,* where choices will be highly reactive to the then-current circumstances the patient finds himself or herself in. The timing of such choices might always be revised in consultation with the physician; but what would be culturally reinforced would be the general commitment to advance planning for one's own death as well as a commitment to assuming a comparatively autonomist, directive role in it. Self-directed dying would be the norm, though of course different people would direct their deaths in quite different ways.

If profound changes affecting matters of how we die are already underway—the epidemiological transition, shifting from parasitic and infectious disease deaths to deaths of predictably degenerative disease; the changes in religious conceptions of suicide so that it is not understood primarily as sin; and the steadily increasing attention to patients' and terminally ill patients' rights of self-determination—it is an open conjecture whether this is where we may be going. Are we in fact experiencing just a temporary aberration in our basic cultural patterns of death and dying, an aberration which is a function of the discrepant development of technologies for life prolongation and for pain control? Or are we seeing the first breaking waves of a sea-change from one perspective on death and dying to another, a far more autonomist and directive one much as we have seen changes in reproduction?

Obviously, I can't say. But I can say that if this is what is happening, the assumption that physician-assisted suicide would or should be rare, an assumption still held by both sides in the current debate, will collapse. We would have no reason to assume that assisted dying should be rare, whatever the relationship between capacities for life prolongation and pain control. Of course, such a picture is very difficult to envision, since we do not think that way about death and dying now. But if we can at least see what is different about viewing personal choices about one's own death *sub specie aeternitatis* and in our current way, *in medias res,* enmeshed in particular circumstances, we can understand why it might occur.

Would it be a good thing, or a bad thing? I can hardly answer that question here, but let me close with a story I heard somewhere in the Netherlands several years ago. I do not remember the exact source of the story nor the specific dates or names, and it is certainly not representative of current practice in Holland. But it was told to me as a true story, and it went something like this:

> Two friends, old sailing buddies, are planning a sailing trip in the North Sea in the summer. It is late February now, and they are discussing possible dates.
>
> "How about July 21?" says Willem. "The North Sea will be calm, the moon bright, and there's a music festival on the southern coast of Denmark we could visit."
>
> "Sounds great," answers Joost. "I'd love to get to the music festival. But I can't be gone then; the twenty-first is the date of my father's death."
>
> "Oh, I'm so sorry, Joost," Willem replies. "I knew your father was ill. Very ill, with cancer. But I didn't realize he had died."
>
> "He hasn't," Joost replies. "That's the day he will be dying. He's picked a date and made up his mind, and we all want to be there with him."

Such a story seems just that, a story, a fiction, somehow horrifying and also somehow liberating, but in any case virtually inconceivable to us. But it was not told to me as a fiction, but as a true story. I've tried to explore the conceptual assumptions that might lie behind such a story, and to consider whether in the future such stories might become more and more the norm. I have not tried to say whether this would be good or bad, but only that this might well be where we are going. In fact, I think it would be good—just as I think increasing personal control over reproduction is good—but I haven't argued for that view here.

In this respect, what the Supreme Court has done in its 1997 *Washington* v. *Glucksberg* and *Vacco* v. *Quill* decision may make a substantial difference. To ban physician-assisted suicide altogether would have been to reinforce the conception that physician-assisted suicide, if it occurs at all, should be rare; to recognize it as a constitutionally based right would have been to begin to create the psychological and legal space within which individuals could reflect in a longer term way about their own future choices when they come to dying, perhaps making physician-assisted suicide an eventual part of their plans, and indeed planning whatever family gatherings, ceremonies, and religious observances they might wish—not as a desperate last resort or reactive escape from bad circumstances, but as a preemptively prudent, significant, culminative experience. To leave the matter with the states, as the Supreme Court in fact did, will be somewhere in between, depending on whether most individual states respond by reinforcing prohibitions or permitting legalization; here, only time will tell.

DISCUSSION QUESTIONS

1. Do people always have the right to choose when to die, only when they have a medical condition that limits their quality of life, or never?

2. What constitutes harm? Can living be a greater harm than death?

3. Should doctors, who dedicate themselves to saving lives, play an active role in taking life?

4. How can trust be damaged by physician participation in euthanasia? Could harm to trust be caused by physicians refusing to aid in these moments?

5. How could PAS be abused?

6. What would be appropriate regulations for PAS or euthanasia (or should there be any)?

7. Is there any point at which a person should be required to die?

8. Is quality or quantity of life more important when making these choices? How can policy reflect that?

9. What does it mean to declare someone dead?

Recommended Cases

Brittany Maynard—Having been diagnosed with brain cancer, Brittany Maynard and her husband went from California to Oregon, where patients have the right to die, so she could die through physician-assisted suicide. The videos she made with her family and friends advocating the expansion of this right have millions of views.

Dr. Death—Jack Kevorkian was a physician who aided more than 100 patients to end their lives via physician-assisted suicide. He did this with prescriptions until the state of Michigan took away his medical license. After this, he was able to allow patients to use a machine he created that allowed drugs to bring about their death through an IV, though this was no longer able to be done in a medical facility. After aiding Thomas Youk in his death, one in which he had to euthanize Youk due to Youk's paralysis from ALS, Dr. Kevorkian was tried for second-degree murder. The judge instructed the jurors that, because this was a murder case, whether Youk consented in his death was irrelevant. Kevorkian was *convicted* and served eight years in prison.

Dax Cowart—In 1973, Dax Cowart was injured a car explosion that killed his father. After the explosion, Cowart was covered in extreme burns (the only places that did not suffer burns were the bottoms of his feet) that caused constant and extreme pain. The doctors who treated him said he was not competent to refuse treatment because of the pain he was in. A psychiatrist said he had decision-making capacity, but the doctors refused Cowart's refusal. After being forced into months and years of painful treatment he did not want, he was able to sustain himself and became a successful lawyer. While he enjoyed his life and did not kill himself, he still believed the doctors should not have forced him to live. He died in 2019 at the age of 71.

UNIT VII

Genes and Data

....

Unit VII Introduction

In 1590, the father-son duo of Hans and Zacharias Janssen created the first microscope. Their invention changed the understanding of our world: a new realm of knowledge was opened. In 1675, Anton van Leeuwenhoek observed bacteria, forever altering our understanding of health. In 1677, Nicolaas Hartsoeker observed sperm under a microscope and posited homunculus living in each individual sperm. Drawings of the homunculus in sperm created a new, and false, understanding of reproduction and masturbation, both of which came with dictums based on this scientific revision. Though germ theory had been proposed hundreds of years earlier, observations at the macro and micro levels were able to progress science beyond miasma and bile theories to become the accepted theory of disease.

New technologies raise new fears, hopes, and questions. Balancing fears and hopes to answer the difficult question of "What should we do with this?" requires an understanding of the technological, medical, and ethical information available. The value of stem cells, and the ways in which they can be used, caused discomfort; artificial intelligence, a tool able to increase medical empathy or discrimination in treatment, has woven itself into the fabric of medicine; big data, the catchphrase of the day, allows for personalization and loss of personhood. CRISPR/cas9, a gene-editing technology that has altered the genetic makeup of animals, plants, and two children, has raised ethical concerns from all who have encountered its use. This unit focuses on understanding the ethical issues that arise with technological progress.

The position of the bioethicist, and the role of bioethics as it relates to technology and the public, is demonstrated in "Expert Bioethics as Professional Discourse: The Case of Stem Cells." By examining the relationship between public policy and bioethics, we see the uses of language and rhetoric that necessitate adaptation by the bioethicist. We examine a variety of political issues as they relate to stem cells, such as federal funding research related to abortion and cloning, to examine the role of bioethics in public debate.

"Keep CRISPR Safe: Regulating a Genetic Revolution" argues for regulation of CRISPR/cas-9 by bioethicists and the scientific community. Their argument is twofold: first, to limit risks and conduct responsible research, and second, to prevent governmental interference that could limit progress. We learn of the potential damages and hopes that gene-editing can bring while also encountering a proposal for its responsible use.

In "Big Data in Medicine: Potential, Reality, and Implications," we learn about RCT and EMR data, the building blocks of big data. This data is the basis for many medical and public health decisions, but also poses ethical concerns, such as bias, low quality of data, and privacy concerns. The tension among these issues and the widespread use of big data is explored.

"Loopthink: A Limitation of Medical Artificial Intelligence" discusses artificial intelligence's role in making difficult ethical choices. As artificial intelligence advances in imitation of our thoughts and behavior, it may also be imitating our worst qualities, such as groupthink and stereotyping. Unable to come to terms with some humane notions, such as dignity, artificial intelligence may have limitations in bioethics.

"Expert Bioethics" as Professional Discourse

The Case of Stem Cells

Paul Root Wolpe and Glenn McGee

P ublic policy debates are exercises in rhetoric. The first battle is often a struggle over definitions, and the winning side is usually the one most able to capture rhetorical primacy by having its definitions of the situation accepted as the taken-for-granted landscape on which the rest of the game must be staged. Public debates, however, are not played out on neutral turf. Players make alliances, exercise power, make claims of legitimacy through expertise, and struggle to gain the cultural and political authority to have their perspectives written into policy directives and law. Powerful public movements, such as recent grass-roots opposition in Europe to genetically modified foods, show how large-scale public resistance can recast the debate in terms other than those defined by scientific, academic, or commercial experts.

Research using human embryos, parts of embryos, potential embryos, transgenic embryos, blastocysts and fetal tissue containing primordial cells, and cloned embryos or cloned transgenic embryonic cells—collectively referred to as human embryonic stem (hES) cell research—has been controversial since the identification of the human pluripotent embryonic stem cell in 1998. The hES case is an interesting one because the general impression is that there has been open public dialogue on the issue. In reality the debate has been one among elites who largely managed to shepherd the controversy quickly toward foregone principles and conclusions in which many of the involved experts were invested. In addition, using formal bodies such as the National Bioethics Advisory Commission (NBAC) or the American Academy for the Advancement of Science (AAAS) to consider the matter of hES cell research gives the impression of carefully considered public dialogue.

In fact, public debate has been minimal, and the formal bodies that considered the issue are themselves closely connected to the research federations and institutions that are pressing for the acceptability of hES cell research. The debate was conducted predominantly under a rubric that we term *expert* bioethics, in which issues are framed and conceptualized at a high level of academic sophistication and political authority by groups of highly skilled professionals who are deputized to identify and resolve moral conflict. Most of them are invested in the medical-industrial complex that is vested on one side of the debate. Even NBAC was appointed at the leisure of the sitting president who created that body and who, in his letters charging NBAC with its tasks, made clear that he opposed cloning but favored hES cell research. Twice NBAC issued official opinions that mirror the concerns and opinions of the president.

Reports of the hES controversy in major scientific journals show how the terminology and conceptual framing of the debate by experts are narrow and reflect the concerns of a small, professionally invested elite. Although an appropriate public language to frame the debate is not developed, we believe that a broader and potentially volatile public dialogue is inevitable. We therefore advocate a shift away from public investment in expert bioethics and toward expert-assisted grass-roots debate and discussion aimed at developing, first and foremost, an appropriate and honest public language for the discussion.

One reason that hES cell research, unlike cloning, has not been taken up in any large-scale way by the public media is the success its defenders had in defining the technology early in its development in ways that made opposition and public debate more complicated and difficult to mount. The research was likened at different times by its critics to cloning, to research on human embryos, and to abortion. At each stage, the professional debate began not over the ethical valence of the objection but over the appropriateness of definitions or categorizations themselves. Both supporters and opponents understood a fundamental principle of the politics of rhetoric: whoever captured the definition of hES cell research had won half the battle. If the debate could be configured as being over abortion, including the ban on using human embryos in federally funded research, opponents would in some sense have won. If, on the other hand, supporters could distance the rhetoric from such concerns, they would gain the upper hand.

Ultimately, none of the attempts of opponents to cast the debate in terms that questioned or attacked hES cell research were successful. They could not overcome the considerable authority of the supporters—the prestige of those framing the debate, their institutional legitimacy, and, perhaps most important, their greater access to professional journals whose commentaries and interpretations of the issue informed the lay media. The lay media reported the controversy, but it was not given the profile of other controversies such as cloning, nor were underlying ideologic struggles clearly articulated for the public. The debate was engaged forcefully in journals such as *Science* and *Nature*, but these journals were overt and covert participants in the attempt of researchers to wield scientific expertise as a weapon to control definitions.

Rhetorical Strategies

Early discussion was mostly limited to the relative obscurity of scientific journals. A cursory glance at relevant articles might lead the lay reader to conclude that the investigation of hES cell research is a technical and scientific matter, or at best one intriguing clue in the larger debate about genetic technology. Yet the debate soon became a political issue, culminating with a charge by the president to NBAC to make a policy recommendation. Politicians, major scientific associations and journals, special interest groups, scientists, and bioethicists became involved.

Cloning

The hES cell research came on the heels of the controversy over cloning, and the early debate was not so much about abortion as about cloning. The question arose as to whether the research might violate state bans on human cloning or the Food and Drug Administration (FDA) memorandum forbidding clinical use of human cloning technology. Virtually all the popular media reports (and some scientific media as well) quickly accepted the analogy between hES cell derivation and cloning, and often refered to the former as a "form of cloning." Politicians who oppose embryo research quickly pointed to a slippery slope between them. Congressman Jay Dickey (R.-Ark.) told *Nature*, "There are no instances in which I feel the ban on federally funded research on human embryos should be lifted. The language of the ban prevents taxpayer funding for bizarre experiments, such as cloning. Eventually, I could see the embryonic stem cell technology going in this direction" (Butler 1998).

Gun shy from the drubbing taken by the scientific community over cloning, the journal *Nature* wanted to make the difference very clear: "... to describe research using human hES under the generic and emotional description of human cloning, as some reporters continue to do, muddies the waters unnecessarily" (editorial in *Nature* 1999). The first argument over who would control how we talk about hES cell research had been engaged; *Nature* wanted to distance these cells from the term "cloning" to insulate the research from the emotional valence of the cloning debate. Yet the most promising hES cell research involves nuclear transfer from somatic cells to enucleated oocytes. This clearly constitutes an important test of the viability of one part of reproductive human cloning, and embryos produced in that way might eventually be gestated as clone offspring. Moreover, many of the ethical issues, under close inspection, have close moral and technologic parallels to those posed by the science or eventual practice of human cloning. In that sense the controversy over hES cell research would benefit from a comparison with the ethical propositions made in such depth and under such close scrutiny throughout the cloning debates. Hence differentiating hES cells from cloned cells was an important move by its supporters to avoid becoming embroiled in the human cloning issue.

Pluripotence versus Totipotence

The journal *Science* noted the public confusion between cloning and hES cells obtained through nuclear transfer; however, the latter is not cloning because the resultant cells are not necessarily totipotent (Solter and Gearhart 1999). The degree to which they may be totipotent is still uncertain and depends on definitional technicalities. The derivation of hES cells and development of nuclear transfer technology cast doubt over what words such as totipotency and pluripotency really mean, and whether their definitions should be modified in light of new conventions about the power of DNA in cells of various kinds and stages of development. However, *Science* maintained that an hES cell derived from a blastocyst of any kind is not totipotent because it cannot be *directly* implanted into a uterus and grown into a conceptus. It may not be too long before that is no longer the working definition of totipotence.

The importance of the issue of totipotence was underlined by Michael West, founder of ACT, a company that claimed to have produced a hybrid cell where human nuclei were inserted into the ova of cows. The hybrid cells were justified by ACT because, unlike cloned cells, they are not totipotent and thus are more like cells than embryos. West thus concluded that the combination of cow and human was not a moral issue at all since "these cells cannot become human beings" (Alper 1999). He further claimed that when nuclear transfer is used to make the embryonic source of hES cells, the resulting embryo is itself not really a human embryo, because its creation does not involve either conception or, in the case of the cow ova, enough human parts.

The debate about totipotency and pluripotency is carried on by scientists with authority and clarity that suggest that well-understood standards exist for such matters. Many of those working on the research insist that existing definitions of the power and properties of cells are simply a matter of objective observation, and that making reference to these definitions can resolve or at least clarify associated moral issues. But no real clarity is found in the history of scientific study of the embryo, or even in current canonical works on the properties of embryos and cells. Discovery of the hES cell and research into its powers and properties—as well as those of cells that are created or hybridized using hES cells—just raise more questions. The most plain conclusion to be drawn from debates about totipotency, pluripotency, and the power of hES research to create clones or embryos is that society has begun to make up new definitions for the powers of cells (McGee and Caplan 1999a). The process of deciding who will refine, reform, or reify definitions of these cells is a sociomoral exercise that has implications for the broader battle for or against hES cell research.

Abortion and Federal Funding

By denying the relationship between hES cell research and cloning, experts managed to avoid emotionally charged cloning rhetoric. However, the more serious charge was that hES cell research required destruction of embryos, and violated the ban on use of embryonic tissue in federally funded scientific investigation. The ban was closely tied to the abortion debate, a debate in which the scientific community was also loath to engage. If hES cells were identified in the public mind as embryos or even as totipotent cells, great pressure would be applied to invoke the ban. The effect of such a measure would be to stifle federal support for hES cell research, which would increase both the burden on researchers to generate quick clinical applications and intellectual property, and the stigma associated with researchers who would be "complicit" in destroying embryos (Robertson 1999). "Embryo research" as a moniker would make it far more difficult for these researchers to enlist the support and funding of patient advocacy groups who might straddle the fence on abortion. [In early August 1999, the American Cancer Society (ACS) withdrew from Patient's Coalition for Urgent Research (Patient's CURe), a coalition of more than thirty patient groups that lobbied Congress to support hES cell research, due to pressure on ACS from officials of the Roman Catholic Church.]

To head off such objections, the scientific community was quick to claim that hES cells are not themselves embryos and so research on them does not violate the law. Bioethicists, including one of the authors (GM) participated in that effort at public hearings held by Patient's CURe in government testimony, and in the professional literature (McGee and Caplan 1999a,b). In January 1999 the U.S. Department of Health and Human Services (DHHS) issued a legal opinion saying that hES cell research does not fall under the federal ban on human embryo research because such cells do not constitute an "organism" as described in the legislation. As long as researchers did not themselves destroy embryos, the cells could be used, because "even if the cells are derived from a human embryo, they are not themselves a human embryo" (Wadman 1999a).

Yet, even if the cells were not embryos, their sources were. The DHHS opinion involves sleight-of-hand to the effect that because the researchers did not *extract* the cells from embryos, they are not considered engaged in embryo research. Frank Young, a commissioner in the Food and Drug Administration under President Reagan, pointed out the strange kind of reasoning the scientific community was employing: "To say, on the one hand, that you cannot support the deliberate destruction of living human embryos to harvest their stem cells, but that you will, on the other hand, pour millions of taxpayer dollars in support of research that you know can only take place using materials derived from that destruction, is an exercise in sophistry, not ethics" (Wadman 1999b). Michael West also made the point that researchers in his group were not destroying embryos when they made and then destroyed cow-human embyro-like organisms, because a cow-human "embryonic combination" is not viable and thus, they claimed, cannot be called an embryo.

Recognizing the difficulty of arguing that research on hES cells derived from destroyed human embryos does not support that destruction, Arthur Caplan hold that using the term "embryo research" to refer to work in which embryos used are donated leftovers is a misnomer. "The embryos existed, were not intended to be used, and were not created for the purpose of research" (Brower 1999, 140). Researchers removed cells from donated embryos and thus Caplan did not regard them as viable embryos. "They were not created for the purpose of research, which would be both forbidden by US law and morally objectionable." In other words, embryos that are leftover, that is, not intended for implantation, are by definition nonviable and therefore are not embryos that would fall under the ban. Many would disagree that human embryos cease to be embryos simply because no one wants to impant them. But, more important, the view that because the embryos were not created for research they can be viewed as cell donors does not necessarily follow. Moreover the fact that donated embryo are destroyed and are nonvoluntarily donated surely complicates donation in the view of those who object to the research on the grounds that embryos are vulnerable human subjects that should not be experimented on.

One proposal that gets around this objection is that embryos, even if they are vulnerable subjects and even if they exist as full moral persons, are not destroyed in the process of hES cell derivation. Because a blastocyst's genetic information is not destroyed in derivation or subsequent use of the cells, recreation of the original donor embryo might be possible using nuclear transfer of recovered, recultivated, donated cells into an enucleated egg (McGee and Caplan 1999b). This was proposed as an illustration of how difficult it is to identify the destruction of embryos in hES cell research, and its authors concede that recreation of a destroyed embryo in no way diminishes the violence involved in transferring DNA from an original embryo into a new cell. Still, the example has been cited widely by experts as evidence that ethics are on the side of mining particular embryos and embryonic tissue, however irrevocably, for hES cells. The definition of "killing" was altered to exclude destruction of potentially reproducible embryonic individuals, eliminating the moral problem in view of research supporters.

The Politics of hES Cell Research Rhetoric

Guided by interested scientists and ethicists, scientific media steered the debate carefully down the middle road and defined hES cells out of all problematic categories: they were not embryos, they were not cloned cells, they were not totipotent. They were simply cells. The media never managed to get a handle on why research ethics were important, or to mobilize the public into a meaningful discussion of the issue. Despite the work of antiabortion interests in the Roman Catholic Church, who enlisted their most public intellectuals to fight the battle, hES cells never became a significant national issue.

In the meantime, bodies that were charged with making policy or advising policy makers on the issue were composed overwhelmingly of scientific elites and experts drawn from the research community. The president asked NBAC to produce a quick report and policy recommendations. Harold Varmus, then director of the National Institutes of Health (NIH), noted that he did not have to wait for NBAC's report for NIH to make its decision (Marshal 1998a). Given the total lack of legislative or regulatory power allocated to NBAC, and its rubber-stamp report that did not change law or policy about cloning, it was a foregone conclusion that little could come of NBAC's expert recommendations. As noted above, few in Congress or the media doubted the outcome of NBAC's deliberations, given the scope of its task and the context of its appointment. The AAAS also put together a panel to discuss and make recommendations on hES cell research ethics, and issued a report that mirrored, no surprise, the editorial calls of *Science* for the research.

Special interest maneuvering was going on behind the scenes. In response to a plea from the House pro-life caucus, a group of seventy congressmen and seven senators wrote a letter to Donna Shalala, Secretary of DHSS, protesting the department's decision to fund hES cell research. In response, a group of thirty-three Nobel laureates wrote to President Clinton and the U.S. Congress in March of 1999 to urge the government to permit such research.

But the real work on establishing the legitimacy of the research was being done in the federal agencies tied to research interests. The use of elites to give the imprimatur of ethical review on research has become *de rigeur* in science. Wary of the controversy, Varmus proposed that an outside committee of experts review all hES cell-related grant proposals to square them with the criteria by Congress (Marshall 1999). When Michael West was asked why he released preliminary results on his bovine-human cell hybrid to the media without the full scientific data that would prove to skeptics that it was in fact such a fusion, he replied that he wanted to develop the technology, but was worried about the reaction of the public and the applicable law. "So I decided," he remarked, "let's talk about the preliminary results. Let's get NBAC to help clear the air." Despite its lack of legislative or regulatory power, in other words, NBAC came to be seen by many in the scientific community as the de facto ethical deliberative body for controversial scientific research (Marshall 1998b).

Other countries are also using deliberative bodies for a similar purpose. In Germany, it is illegal to use hES cells from spare eggs or embryos, as the Embryo Protection Law confers full human rights from the moment of conception. The law includes cloned embryos as well. However, there is no law against developing pluripotent human embryonic germ cells from aborted fetuses. Germany's basic research funding agency, Deutsche Forschungsgemeinschaft, called for establishment of a central committee to assess the issue, as well as open dialogue at the European level (Abbot 1999).

In Britain, the government receives recommendations on these issues from the Human Genetics Advisory Commission and the Human Fertilisation and Embryology Authority. These groups sought public comment through one of their public "consultations," increasingly part of British debate about bioethics, and found that whereas 86 percent of people who commented were against human reproductive cloning, only a fraction wanted to limit hES cell research, and the recommendation of the two commissions reflected these positions (Marshall 1998c). Nevertheless, in a bow to anticipated public debate, Britain imposed a temporary moratorium on certain types of hES cell research. It already issues licenses for those wanting to do research on embryos in the first fourteen days of development (a limited number of such experiments are permitted), and the advisory commission recommended a similar strategy of research licenses for cloning and hES cell research. France's National Bioethics Committee recommended loosening the ban on research on human embryos to allow cultivation of hES cells (Butler 1997).

Private industry is using the same tactics to preempt ethical challenges. Worried about public opinion on their hES cell work, Geron engaged its own ethics board to make recommendations.

Smithkline Beecham, one of many large pharmaceutical corporations likely to do hES cell research, held special ethics hearings in which it too searched for a way to find acceptable language with which to promote this work.

Who Should Make Bioethical Decisions?

The public debate about hES cell research reflects an odd shift from the equally problematic debate about cloning. Regarding cloning, experts struggled to keep up with an expanding maelstrom of public fear and expressions of anger. Institutions were asked by their contituents what their position was and what actions they were going to take. In contrast, the political system and its participants at various levels were ready for the hES cell debate, and the last thing they wanted was another Dolly controversy. Public understanding of the phenonemon must be prefigured and the parameters of the debate defined before the science becomes well known and widespread. The lesson gleaned by the experts from the cloning controversy was simple: *in modern biotechnologic controversies, public debate must be shepherded and fostered by an elite that is prepared to seize rhetorical primacy, and to mold existing institutions or create new ones for that purpose.*

The result of the cloning debate on bioethics as a whole (it is not too early to say) has been paradoxical. After cloning, bioethics plays a more important role than ever in the discussion of science; yet its role has been largely assimilated into the political model of Camelot, in which the philosopher is further elevated above the people and the model of discourse further rarified. The result is NBAC, whose decisions are based on the testimony of elites (many of whom have interests in the medical-biotechnical enterprise), whose membership represents overwhelmingly one side of the politicocultural spectrum, and whose publications cannot be understood by more than a fraction of the public.

If the current model of expert bioethics, somewhat amplified by the cloning controversy, is not desirable, what is the proper role of the bioethicist in matters public? What is the task of the philosopher, social scientist, or clinician with skills in moral matters? The hES cells in all their complexity provide, we believe, an obvious opportunity to test a new model for the role of bioethics in public debate. Before this can happen several key claims must be accepted by elites in the field. First, although bioethics as a discipline prides itself on its openness to public participation, few participants have any mastery whatever of the implements of public debate, including rhetoric, media, and politics. Bioethics must grow more familiar with these skills, while avoiding the sale of its soul or similar distortion of purpose. Second, the role of language in public debates about ethics is an important one, and skills of discerning moral and scientific valences of particular words are crucial. The effort to frame hES cell research in clear language that neither prejudices unnecessarily nor misleads is the most important moral effort now under way, and the role of the bioethicist should be as critic of, and not handmaiden to, expert efforts to smuggle values into the process. Bioethics can be the proctor of public debate only if it plays this explanatory role with more skill of discernment: rather than accepting the debate at face value, we need a bioethics that questions the false or patently political framing put to NBAC or other bodies of discussion and debate. Third and most important, bioethics must be a public activity itself, written more and more with a broad audience in mind. Even where this goal undermines the first and second goals, it should be reflected in a commitment of bioethics mentors to educate a new generation of scholars whose ambitions for writing and thinking extend beyond a small group of philosophical or social scientific peers.

References

Abbott, A. 1999. Don't try to change embryo research law. *Nature* 398: 275.

Alper, J. 1999. A man in a hurry. *Science* 283: 1434–1435.

Brower, V. 1999. Human ES cells: Can you build a business around them? *Nature Biotechnology* (Feb. 17): 139–142.

Butler, D. 1997. France is urged to loosen ban on embryo research. *Nature* 387: 218.

_____. 1998. Breakthrough stirs US embryo debate. *Nature* 396: 104.

Editorial. 1999. Towards the acceptance of embryo stem-cell therapies. *Nature* 397: 279.

Marshall, E. 1998a. Use of the stem cells still legally murky, but hearing offers hope. *Science* 282: 1962–1963.

_____. 1998b. Claim of human-cow embryo greeted with skepticism. *Science* 282: 1390–1391.

_____. 1998c. Britian urged to expand embryo studies. *Science* 282: 2167–2168.

_____. 1999. NIH plans ethics review of proposals. *Science* 284: 413–415.

McGee, G. and Caplan, A. 1999a. What's in the dish? Ethical issues in stem cell research. *Hastings Center Report* 29(2): 36–38.

_____. 1999b. The ethics and politics of small sacrifices in stem cell research. *Kennedy Institute of Ethics Journal* 9(2): 151–165.

Robertson, J. 1999. Ethics and policy in embryonic stem cell research. *Kennedy Institute of Ethics Journal* 9(2): 109–136.

Solter, D. and Gearhart, J. 1999. Putting stem cells to work. *Science* 283: 1468–1470.

Wadman, M. 1999a. Embryonic stem-cell research exempt from ban, NIH is told. *Nature* 397: 185–186.

_____. 1999b. White House cool on obtaining human embryonic stem cells. *Nature* 400: 301.

Keep CRISPR Safe

Regulating a Genetic Revolution

Amy Gutmann and Jonathan D. Moreno

The possibility of rewriting the genome of an organism, or even of an entire species, has long been the stuff of science fiction. But with the development of CRISPR (which stands for "clustered regularly interspaced short palindromic repeats"), a method for editing DNA far more precisely and efficiently than was possible with older technologies, fiction has edged closer to reality. CRISPR exploits an ancient system that allows bacteria to acquire immunity from viruses. It uses an enzyme called Cas9 to cut strands of DNA at precisely targeted locations, allowing researchers to insert new genetic material into the gap.

CRISPR promises to revolutionize gene editing, which comprises two distinct but related fields. The first involves a technique to modify inherited genes in nonhuman organisms in order to spread a trait throughout a population, using a process known as a gene drive. The other involves editing the human genome, either in normal body cells (known as somatic cells) or in the germline, the cells that pass genes down to offspring.

The advances made possible by CRISPR could bring vast benefits to society, but the technology also poses risks. An out-of-control gene drive could drastically alter or even threaten a species. And editing the human genome raises risks both for individuals and for society as a whole. To head off those dangers, governments and scientific institutions will have to respond by establishing standards that both enable promising research to go forward and reassure the public that the work is being conducted responsibly. Yet especially when the science is at such an early stage, there is a risk that governments will do too much rather than too little. To avoid that problem, the global scientific and biological ethics communities must take the lead, designing standards and procedures that reduce the dangers of these powerful new technologies without forgoing the benefits.

Playing God

The goal of a gene drive is to spread or suppress certain genes in a wild population of organisms. It works by exploiting a quirk of nature. In sexually reproducing species, most genes have a 50 percent chance of being passed from parent to child, as offspring receive half their genes from each parent. As a result, genetic mutations normally spread only if they make an organism more likely to survive or breed. But some genes have evolved mechanisms that give them better than 50 percent odds of being passed on. That allows changes in those genes to proliferate quickly even if they have no effect on evolutionary fitness. Scientists can exploit this tendency by using CRISPR to insert genetic material into the "selfish" part of an organism's genome, ensuring that the new trait will be passed on to most offspring, eventually spreading through large populations.

This process could be exploited both to improve public health and to promote economic development. Scientists could use gene drives to break disease transmission chains, eliminating the need to use costly and harmful insecticides. For example, researchers are looking at using the technology to inhibit the transmission of Lyme bacteria from mice to ticks, a move that could wipe out Lyme disease among humans, since humans can catch the disease only from tick bites. In agriculture, gene drives could immunize plants against many kinds of pathogens and curb or eliminate populations of invasive animals, such as mice, that destroy crops. Researchers are working to develop a house mouse that can give birth only to male offspring. If the altered mice were released into the wild, female mice would gradually disappear and the species would die off within the area in which the altered mice were introduced.

Gene drives could also reverse some worrying environmental trends. Many amphibian species, such as frogs, toads, and newts, have suffered catastrophic declines over the last few decades. If scientists introduced engineered genes that rendered amphibians immune to common pathogens, many species could recover.

But alongside these benefits come serious risks. A gene drive gone wrong could leave a species extinct or introduce dramatic and unintended effects. Most gene drives would likely be limited to a single place, such as an island or an isolated and containable area of land, at least at first. If a genetically modified animal or plant escaped, however, then the gene drive could spread uncontrollably. And if a modified organism mated with a member of another species, it could transmit the changes to new populations. Entire species could be wiped out and ecosystems upended.

These risks are dramatic, but they are also far off. In part because the research is at such an early stage, scientists disagree about how much and what kind of regulation and guidance will be required. So governments should follow the principle of regulatory parsimony, which dictates that they should impose only those restrictions necessary to maintain ethical standards and public safety. Doing so will maximize the scope of free scientific discovery in a way that is consistent with serving the public good. In the countries where much of the research is taking place, before imposing new restrictions, governments should rationalize the current system of regulation. In the

United States, for example, the Department of Agriculture, the Food and Drug Administration, and the Environmental Protection Agency all share responsibility for different aspects of field trials and commercial products. They should harmonize their biological safety guidelines for gene-drive studies. On the international stage, the Cartagena Protocol on Biosafety, an agreement among most of the world's countries that came into force in 2003, regulates genetically modified organisms, but the United States has not signed it.

In lieu of formal regulations on gene drives, scientists could agree to build safety measures into gene-drive systems, such as alterations that would cancel out previous drives or gene modifications designed to grow less frequent over time, so that successive generations would express the gene less and less once the original problem has been sufficiently ameliorated. Researchers will also need to be transparent about their work and consult local communities to gain consent before introducing gene drives into the wild.

Humans 2.0

Just as challenging for the global scientific community will be the issues raised by human genome editing. A wide range of diseases have been identified as potential targets for treatments that modify genes in a patient's somatic cells, including certain cancers, cystic fibrosis, hemophilia, HIV/AIDS, Huntington's disease, muscular dystrophy, some neurodegenerative diseases, and sickle cell anemia. Developing therapies for these conditions will not be straightforward. Preliminary laboratory research suggests that the human immune system may resist the version of the enzyme Cas9 currently used in CRISPR. If that result holds up, either Cas9 will have to be modified or replacement enzymes will have to be developed. Yet this does not appear to be a major setback, as scientists are already using other enzymes for techniques associated with CRISPR.

Whatever form it takes, research involving human gene editing will have to meet the already rigorous regulatory standards governing medical research. In the United States, all applications for clinical experiments must be approved by the Food and Drug Administration and reviewed by the National Institutes of Health. The FDA advises researchers to follow up with participants in gene therapy trials for as long as 15 years after the end of the trial to discover and deal with any delayed ill effects. And once the FDA approves a gene therapy product for public sale, it requires companies to monitor its use, report any adverse events, and give public warnings as appropriate. This regulatory regime will be sufficient when it comes to using CRISPR to edit somatic cell genes given that the process, although different from other techniques for modifying cells, does not raise new safety or ethical issues.

The same cannot be said for interventions that modify an individual's germline, which carries genes that are inherited. Such interventions raise both the prospect of vast benefits and thorny questions of safety and ethics. Gene editing could, in theory, prevent the transmission of genes that increase the risks of life-threatening diseases, such as breast cancer or cystic fibrosis. Families with histories of breast cancer associated with certain mutations in the BRCA1 and BRCA2 genes (which help prevent tumors), for example, may wish to protect their descendants by editing their genes.

More speculative are germline modifications intended to make future children stronger, better looking, or smarter. The prospect of such genetic engineering raises the specter of disastrous twentieth-century experiments in eugenics, although today most of the demand would likely come from individuals rather than states. To the extent that state projects did attempt to enhance national populations, they would be ill advised and socially disruptive. What is more, because of the enormous complexity of traits such as intelligence, the results of those projects would certainly disappoint their proponents.

Reengineering the human genome raises risks not only for individual patients but also for humanity as a whole. Unlike the generations of rapidly propagating species, such as mosquitoes, human generations span many years, so any harmful change in a human germline could take decades or even centuries to become pronounced. But that does not mean that the risks should be ignored. Adjusting one part of complex human societies could well have serious consequences for public health, economic growth, and social cohesion.

In 2017, the U.S. National Academies of Sciences, Engineering, and Medicine recommended that researchers exercise caution when it comes to efforts to prevent disease transmission through gene editing but said that such work should be allowed to go forward, albeit under "stringent oversight." The NAS did not extend this recommendation to experiments designed to enhance future generations, which it said should not be allowed "at this time." As the report noted, the risks of enhancement experiments are similar to those of therapeutic ones, but as long as their overall benefits are smaller, they are not worth scarce research dollars.

These sorts of guidelines will shape the work of reputable scientists, but they are not designed to stop rogue actors. At some point, governments may have to pass laws to prevent unscrupulous researchers from abusing gene editing. For now, however, the science is nowhere near advanced enough for policymakers to know what kinds of measures would work.

The Light and the Dark

Experiments in both human genome editing and gene drives are generally classified as "dual use"— research whose results may be used for good or evil. Since the mid-1970s, when scientists developed recombinant DNA technology, which allows DNA from different organisms to be combined to create new genetic sequences that can give organisms new traits, researchers have been concerned about the possibility of bioengineered pathogens, created either deliberately or accidentally. In 2000, researchers in Australia discovered a technique for modifying mousepox that made it more dangerous; this technique could also be applied to smallpox. In 2002, a lab in New York replicated the polio virus using publicly available DNA ordered from a biotechnology company. And in 2012, just as CRISPR was emerging, researchers in the Netherlands and the United Kingdom showed that the wild form of avian flu, which spreads from birds to mammals only through physical contact, could be genetically modified to allow it to move from birds to ferrets—and, by extension, to humans—through the air.

Although gene drives cannot be used to create new viruses or bacteria—neither type of pathogen reproduces sexually—they could be used to create other kinds of weapons. For example, mosquitoes could be modified to produce toxins or such that they can expand their natural habitat and so spread malaria, dengue fever, or other diseases outside tropical areas.

Indeed, virtually all biological research could plausibly be described as dual use. It is hard to think of a major biological breakthrough that could not be exploited for harmful ends. This is one reason why all of those who value the lifesaving breakthroughs that biological research has made possible should reject the idea of regulating biological research primarily because it is dual use. Certainly, if a technology has no conceivable malign application, then regulation should be off the table. But what is equally certain, the mere fact that something can be used to do harm must not suffice to trigger regulation.

At the moment, different countries take widely varying approaches to regulating dual-use research. The United States tends to focus only on select biological agents that threaten public health. The European Union, by contrast, takes a more precautionary approach that requires a risk assessment for any organism that could pose a threat. Regulatory parsimony and a bias toward scientific freedom favor the more focused policy, whereas greater risk aversion favors the precautionary approach.

Self-Government

For all its unprecedented power, CRISPR is of a piece with other research breakthroughs in synthetic biology. It has both enormous potential to transform societies for the better and possible malign uses. Dealing with the latter will require crafting highly specific rules so that regulators don't end up sweeping all CRISPR research into a costly new regulatory net with little or no benefit to society. And even the best-designed regulation cannot eliminate the possibility that researchers will accidentally discover a dangerous new application of a new technology. Before regulators consider additional rules, CRISPR researchers will have to comply with existing scientific norms and regulations, perhaps the field's biggest short-term challenge.

The modern scientific community is both cooperative and competitive. Even so, scientific establishments have shown themselves capable of self-governance when public safety and confidence are at stake. The standards developed by recognized authorities are encouraging: for example, in the first decade of this century, in response to new laboratory practices involving the use of human stem cells in nonhuman animals, national science academies came up with a set of guidelines. The guidelines are voluntary, but they delineate in a well-informed way what is and what is not ethically acceptable, and they have been widely embraced by scientists, the editors of prominent scientific journals, and regulators.

The most effective standards for gene-editing research will come from the scientific community itself, through international summits of science academies and a continual process of intellectual exchange. Those are the forums that can respond best to often unpredictable developments in the science and react sensitively to public opinion. Prudent self-governance among scientists may not produce headlines, but it is the process most likely to enable CRISPR and the next generation of research breakthroughs to reach their full potential.

Selections from "Big Data in Medicine: Potential, Reality, and Implications"

Dan Sholler, Diane E. Bailey, and Julie Rennecker

The Building Blocks of Big Data: RCT and EMR Data

The evidence-based medicine (EBM) movement set the stage for the use of RCT data in big data health care applications (Murdoch and Detsky 2013; Simpao et al. 2014). EBM, broadly construed, is the use of the best-available scientific evidence in diagnosis, prognosis, and treatment decisions (Guyatt et al. 1992). Proponents of EBM argue that RCT data, once aggregated and analyzed, offer the best opportunities for gains in efficiency, quality, and safety in health care (Sackett et al. 1996). EBM involves distilling the results from large numbers of RCT studies into systematic reviews—meta-analyses conducted by human reviewers who are increasingly aided by data mining and other computational techniques (e.g., Wallace et al. 2012)—to stay abreast of the best-available evidence. This task is becoming ever more difficult given the dramatic rise in the number of RCT studies. Bastian, Glasziou, and Chalmers (2010) noted that the number of RCT studies in published articles increased from a hundred in 1966 to twenty-seven thousand in 2007, with, as mentioned above, seventy-five new studies and eleven new systematic reviews appearing daily by 2010.

Worse yet, EBM has struggled to leverage research data to provide useful recommendations for clinical interventions. Systematic reviews routinely conclude that despite plentiful RCT data, more data are needed to draw accurate conclusions. For example, Villas Boas and his colleagues (2013) examined systematic reviews in the Cochrane Library (a collection of more than half a dozen medical databases of abstracts, studies, reviews, and evaluations) and reported that only a small number of them had sufficient evidence for clinical interventions. In 2004, these authors had similarly evaluated the conclusions of Cochrane Library systematic reviews and found that 48 percent of them reported insufficient evidence for use in clinical practice (El Dib, Atallah, and Andriolo 2007). In their 2012 paper, based on 2011 data, that percentage had dropped just a few points to 44 percent. The authors concluded that a large number of high-quality RCT studies are needed to provide sufficient evidence for clinical interventions.

The difficulties in employing RCT data in the EBM paradigm to aid medical decision making caused some scholars to wonder if better gains could be achieved using large data sets of clinical data from patient records. Eric B. Larson (2013), for instance, observed that mining huge data sets of routine health care data from a large percentage of the population might yield, in a matter of months, results that once required decades of RCT data collection from selected samples of patients, with more generalizable conclusions. EMR data are rich in information that might aid such analysis; most EMR data consist of quantitative data (e.g., a patient's laboratory results), qualitative data (e.g., text-based documents and demographic information), and transactional data (e.g., records of medication delivery).

Governmental and administrative incentives recently accelerated adoption rates of EMR systems, thereby increasing the available clinical patient data and raising hope for their use in analysis. According to Nicole Szlezák and her colleagues (2014), between 2005 and 2012, the percentage of US physicians using EMRs jumped from 30 to 50 percent; in hospitals, EMR adoption reached 75 percent by 2012. Moreover, a survey by the American Hospital Association showed that adoption of EMR systems doubled between 2009 and 2011, suggesting that this uptake was a direct result of the 2009 HITECH Act, which mandated the "meaningful use" of health information technology in health care practice, with penalties for practices that failed to adopt EMR systems by 2016 (Murdoch and Detsky 2013).

Although EMR data offer the opportunity to avert some of the generalizability issues that plague RCT studies by studying data from the broad population of actual patients rather than the limited samples of experimental studies, they introduce new issues of patient privacy (Murdoch and Detsky 2013; Kayyali, Knott, and Van Kuiken 2013). The US Health Insurance Portability and Accountability Act is among the federal regulations that protect patients' confidentiality; in addition to meeting its requirements, users of EMR data would need to develop standards for how data would be shared, accessed, and used (Szlezák et al. 2014). These issues notwithstanding, this new wealth of clinical patient data, combined with RCT data, has prompted interest in CDSS that combine both forms of data with analytic techniques and presentation tools for use in medical decision making.

CDSS: Combining Big Data with Analytics

With the potential to make large data sets of RCT and EMR data tractable and useful, CDSS are bringing the goals of EBM to fruition, and today constitute the face of big data in medicine. CDSS and associated technologies, such as the one described by Ryan Abbott (this volume), offer multiple tools for analyzing and interpreting big data. In an examination of eleven CDSS (seven vendor-provided and four internally developed systems), Adam Wright and his colleagues (2011) found fifty-three types of CDSS tools. The most common tools across these eleven CDSS involved medication dosing and order support. Other tools concerned point-of-care alerts, displays of relevant information, and workflow support.

Considerable variability exists among CDSS in terms of the tools they include (Mann 2011). Figure 7.1 orders some of the most common tools along a continuum of physician autonomy, with

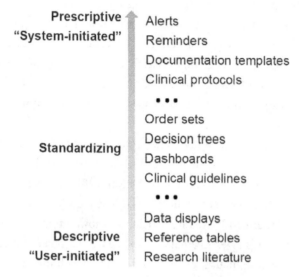

Figure 7.1 CDSS tools along a continuum of physician autonomy

greater autonomy permitted at the bottom of the figure with descriptive, user-initiated tools, and less autonomy at the top of the figure with prescriptive, system-initiated tools.

Beginning at the top of the continuum in figure 7.1, the most prescriptive tools call for users to take specified action or provide a rationale for deviating from a directive. For instance, *alerts* and *reminders*, also called "point-of-care electronic prompts" (Schwann et al. 2011, 869) appear in pop-up-style windows, alerting physicians to errors or reminding them to order periodic screening exams. Both alerts and reminders require user acknowledgment before further action can be taken in the EMR. *Documentation templates* and *clinical protocols* are less intrusive than alerts and reminders, although they are still intended to constrain and direct user action. Documentation templates are simple electronic forms that draw the care team's attention to particular aspects of the care situation (Ash et al. 2012), such as a series of patient-safety precautions to be completed before a surgery begins. Clinical protocols consist of action sequences for managing a specific patient condition, such as postoperative care for hip surgery. The protocols often include multidisciplinary instructions, allowing nurses and allied health professionals to take physician-sanctioned actions without waiting for a direct order.

The standardizing tools are less intrusive than the prescriptive ones, frequently appearing in a sidebar or the appropriate section of the EMR. These tools offer guidance in the form of best practice recommendations. Although these tools are meant to standardize behavior, they do not require user action in the way that prescriptive tools do. *Order sets* include lists of the diagnostic tests, medications, and nursing interventions considered appropriate for patients' diagnoses. *Decision trees*, often in flowchart form using yes/no questions or if-then statements, help physicians make diagnostic or treatment decisions given a patient's circumstances. *Dashboards* give physicians feedback on their decisions and their patients' health status. For instance, a dashboard may show the physician's compliance with a recommended order set relative to local or national averages. Similar to order sets, *clinical guidelines* reflect the current standard of care for a specific diagnosis (Timmermans and Mauck 2005), typically established by one of the professional associations, such as the guidelines for treating congestive heart failure published by the American College of Cardiology. Currently, local organizational and state policies determine how much latitude physicians retain in following or deviating from these clinical guidelines, but shifting payment structures along with increasing calls for physician and hospital performance metrics suggest that application of the guidelines will become increasingly prescriptive (Foote and Town 2007; Miller, Brennan, and Milstein 2009).

Finally, at the descriptive end of the continuum are various reference tools that display information at the physician's request. These tools are often accessed via contextually sensitive hyperlinks or buttons in the EMR, and include *data displays*, which allow a physician to view patient parameters (e.g., lab test results) in graphic form; *reference tables*, such as medication dosages organized by patient weight; and links to the *research literature*.

Many of these CDSS tools raise concerns for their use in practice. Alerts are one such area. Because alerts (a prescriptive tool) disrupt a physician's workflow, they can lead to "alert fatigue," causing practitioners to ignore or override alerts on the grounds if the alerts are too frequent or intrusive (Jenders et al. 2007). As Anurag Gupta, Ali S. Raj, and Ramin Khorasani (2014) observed, physicians may strategically enter patient data incorrectly to avoid frequent alerts. Incomplete or inaccurate information in the CDSS database can prompt incorrect CDSS alerts and reminders, or fail to prompt necessary ones—outcomes that might detract from patient care or increase physician skepticism. As a case in point, Ruchi Tiwari and his colleagues (2013) documented a CDSS failure to issue an interaction alert for a heart transplant patient who subsequently experienced a drug interaction. Afterward, the institution reviewed possible drug-to-drug interactions for the drugs involved. Based on this review, the institution upgraded 62 of 329 possible pairings to more severe alerts in the system. Work continues to determine which drug-to-drug interactions are severe enough to warrant interruption of a physician's workflow via an alert (Phansalkar et al. 2013). Another concern arises from the considerable variability in the sources of data used in order sets and decision trees, two standardizing CDSS tools. For example, in the realm of drug information, free as well as subscription databases exist. These databases differ in their scope, completeness, and ease of use, meaning that CDSS recommendations may vary according to which database the system employs (Clauson et al. 2007).

Overall, customizability appears as a key concern and need for many, if not all, CDSS tools, and may ultimately speak to the success or failure of CDSS. Many researchers note the need for the customizability of alerts and reminders to address patient problems on a case-by-case basis (Jenders et al. 2007). In an interview study of representatives from thirty-four community hospitals, Joan S. Ash and her colleagues (2012) reported that most informants found that the CDSS system needed to be customized much more than they had expected. Customizability may be difficult in systems that provide high levels of specificity; it is also hindered by the considerable variability in the specificity each CDSS supplies. James B. Jones and his colleagues (2013) characterized CDSS specificity as generic or highly tailored based on the number of input variables that generate recommendations. These authors noted that allowing greater specificity exponentially increases the size of the CDSS database, presenting a challenge for managing and curating the knowledge bases, and making the development of actionable guidelines more difficult. Below, we tease out other factors that influence the effective management and use of big data in health care by reviewing reports of CDSS implementation to date. [...].

Future Research

The dramatic rise in CDSS implementation in the past few years has led researchers to explore factors that contribute to physicians' use. Researchers have paid much less attention to the outcomes of CDSS use, however. When researchers do examine outcomes, they tend to restrict their analysis to patient and financial outcomes. These outcomes are important, and deserve continued, improved (as with better metrics), and comprehensive examination. Beyond them, though, lie several other key challenges for researchers in this realm. Based on our review, we consider three challenges in particular here: the need for better data, the likelihood of occupational outcomes, and potential changes in the very nature of medical practice.

Better Data

Critically evaluating the sources of big data in the context of medicine is crucial for developing effective practical applications such as CDSS. For both RCT and EMR data, the quantity and quality of data continue to vex researchers and practitioners. Researchers are only beginning to evaluate potential problems with using RCT data to inform medical decisions. An obvious challenge is developing new ways to collect and analyze the massive amount of data published each year. Additionally, current methods for determining which RCT data should be included in CDSS are yet to account for potential biases in how the data were generated. Machine-learning and natural language processing researchers are developing promising solutions to these problems. For example, Byron C. Wallace et al. (2012) developed novel methods for conducting systematic reviews in genetic medicine and ensuring efficient updating. Likewise, Iain J. Marshall and Joël Kuiper along with Wallace (2014) presented machine-learning methods for assessing the risk of bias in RCT. These researchers have made steady progress in evaluating experimental design and execution as sources of bias, but have yet to assess factors such as who or what funded the study, where it was published, and how the results were reported.

Aside from RCT data, EMR data present their own challenges, mostly associated with patient privacy and methods for extracting data from the record. Patient privacy regulations inhibit data sharing between organizations, limiting the amount of patient data on which CDSS can draw. The implications of privacy concerns for the success of big data applications in medicine are well documented. Less understood, though, are the implications of translating medical data into highly structured formats. Translating data generated from social processes into digitized and computable forms necessitates assumptions as well as decisions on behalf of designers, organizations, and users (Alaimo and Kallinikos, this volume). The assumptions and decisions made by "data collectors" may or may not be aligned with the values of "data subjects" (Burdon and Andrejevic, this volume), or particularly in medicine, the practices of technology users. We explore this potential conflict below. [...].

References

Ash, Joan S., James L. McCormack, Dean F. Sitting, Adam Wright, Carmit McMullen, and David W. Bates. 2012. "Standard Practices for Computerized Clinical Decision Support in Community Hospitals: A National Survey." *Journal of the American Medical Informatics Association* 19 (6): 980–987.

Bastian, Hilda, Paul Glasziou, and Iain Chalmers. 2010. "Seventy-Five Trials and Eleven Systematic Reviews a Day: How Will We Ever Keep Up?" *PLoS Medicine* 7 (9). http://journals.plos.org/plosmedicine/article?id=10.1371/journal.pmed.1000326 (accessed February 10, 2016).

Boas, Paulo José Fortes Villas, Regina Stella Spagnuolo, Amélia Kamegasawa, Leandro Gobbo Braz, Adriana Polachini do Valle, Eliane Chaves Jorge, Hugo Hyung Bok Yoo, Antônio José Maria Cataneo, Ione Corrêa, Fernanda Bono Fukushima, Paulo do Nascimento, Norma Sueli Pinheiro Módolo, Marise Silva Teixeira, Edison Iglesias de Oliveira Vidal, Solange Ramires Daher, and Regina El Dib. 2013. "Systematic Reviews Showed Insufficient Evidence for Clinical Practice in 2004: What about in 2011? The Next Appeal for the Evidence-Based Medicine Age." *Journal of Evaluation in Clinical Practice* 19 (4): 633–637.

Clauson, Kevin A., Wallace A. Marsh, Hyla H. Polen, Matthew J. Seamon, and Blanca I. Ortiz. 2007. "Clinical Decision Support Tools: Analysis of Online Drug Information Databases." *BMC Medical Informatics and Decision Making* 7 (1): 7.

El Dib, Regina P., Álvaro N. Atallah, and Regis B. Andriolo. 2007. "Mapping the Cochrane Evidence for Decision Making in Health Care." *Journal of Evaluation in Clinical Practice* 13 (4): 689–692.

Foote, Susan Bartlett, and Robert J. Town. 2007. "Implementing Evidence-Based Medicine through Medicare Coverage Decisions." *Health Affairs* 26 (6): 1634–1642.

Gupta, Anurag, Ali S. Raja, and Ramin Khorasani. 2014. "Examining Clinical Decision Support Integrity: Is Clinician Self-Reported Data Entry Accurate?" *JAMA: The Journal of the American Medical Informatics Association* 21 (1): 23–26.

Guyatt, Gordon, John Cairns, David Churchill, Deborah Cook, Brian Haynes, Jack Hirsh, Jan Irvine, Gordon Guyatt, MD, MSc, John Cairns, MD, David Churchill, MD, MSc, Deborah Cook, MD, MSc, Brian Haynes, MD, MSc, PhD, Jack Hirsh, MD, Jan Irvine, MD, MSc, Mark Levine, MD, MSc, Mitchell Levine, MD, MSc, Jim Nishikawa, MD, David Sackett, MD, MSc, Patrick Brill-Edwards, MD, Hertzel Gerstein, MD, MSc, Jim Gibson, MD, Roman Jaeschke, MD, MSc, Anthony Kerigan, MD, MSc, Alan Neville, MD, Akbar Panju, MD, Allan Detsky, MD, PhD, Murray Enkin, MD, Pamela Frid, MD, Martha Gerrity, MD, Andreas Laupacis, MD, MSc, Valerie Lawrence, MD, Joel Menard, MD, Virginia Moyer, MD, Cynthia Mulrow, MD, Paul Links, MD, MSc, Andrew Oxman, MD, MSc, Jack Sinclair, MD, and Peter Tugwell, MD, MSc. 1992. "Evidence-Based Medicine: A New Approach to Teaching the Practice of Medicine." *Journal of the American Medical Informatics Association* 268 (17): 2420–2425.

Jenders, Robert A., Jerome A. Osheroff, Dean F. Sittig, Eric A. Pifer, and Jonathan M. Teich. 2007. Recommendations for Clinical Decision Support Deployment: Synthesis of a Roundtable of Medical Directors of Information Systems. In *AMIA Annual Symposium Proceedings*, 359–363. Bethesda, MD: American Medical Informatics Association.

Jones, James B., Walter F. Stewart, Jonathan D. Darer, and Dean F. Sittig. 2013. "Beyond the Threshold: Real-Time Use of Evidence in Practice." *BMC Medical Informatics and Decision Making* 13 (1): 47–60.

Kayyali, Basel, David Knott, and Steve Van Kuiken. 2013. "The Big-Data Revolution in US Health Care: Accelerating Value and Innovation." Insights and Publications. http://www.mckinsey.com/insights/health_systems_and_services/the_big-data_revolution_in_us_health_care (accessed February 10, 2016).

Larson, Eric B. 2013. "Building Trust in the Power of 'Big Data' Research to Serve the Public Good." *JAMA: The Journal of the American Medical Association* 309 (23): 2443–2444.

Mann, Devin M. 2011. "Making Clinical Decision Support More Supportive." *Medical Care* 49 (2): 115–116.

Marshall, Iain J., Joël Kuiper, and Byron C. Wallace. 2014. Automating Risk of Bias Assessment for Clinical Trials." In *Proceedings of the 5th ACM Conference on Bioinformatics, Computational Biology, and Health Informatics*, 88–95. New York: ACM.

Miller, Thomas P., Troyen A. Brennan, and Arnold Milstein. 2009. "How Can We Make More Progress in Measuring Physicians' Performance to Improve the Value of Care?" *Health Affairs* 28 (5): 1429–1437.

Murdoch, Travis B., and Allan S. Detsky. 2013. "The Inevitable Application of Big Data to Health Care." *JAMA: The Journal of the American Medical Association* 309 (13): 1351–1352.

Phansalkar, Shobha, Amrita Desai, Anish Choksi, Eileen Yoshida, John Doole, Melissa Czochanski, Alisha D. Tucker, Blackford Middleton, Douglas Bell, and David W. Bates. 2013. "Criteria for Assessing High-Priority

Drug-Drug Interactions for Clinical Decision Support in Electronic Health Records." *BMC Medical Informatics and Decision Making* 13 (1): 65–76.

Sackett, David L., William Rosenberg, J. A. Gray, R. Brian Haynes, and W. Scott Richardson. 1996. "Evidence Based Medicine: What It Is and What It Isn't." *British Medical Journal* 312 (7023): 71–74.

Schwann, Nanette M., Karen A. Bretz, Sherrine Eid, Terry L. Burger, Deborah Fry, Frederick Ackler, Paul J. Evans, David Romanchuk, Michelle Beck, Anthony J. Ardire, Harry Lukens, and Thomas M. McLoughlin Jr. 2011. "Point-of-Care Electronic Prompts: An Effective Means of Increasing Compliance, Demonstrating Quality, and Improving Outcome." *Anesthesia and Analgesia* 113 (4): 869–876.

Simpao, Allan F., Luis M. Ahumada, Jorge A. Gálvez, and Mohamed A. Rehman. 2014. "A Review of Analytics and Clinical Informatics in Health Care." *Journal of Medical Systems* 38 (4): 1–7.

Szlezák, Nicole, M. Evers, J. Wang, and L. Pérez. 2014. "The Role of Big Data and Advanced Analytics in Drug Discovery, Development, and Commercialization." *Clinical Pharmacology and Therapeutics* 95 (5): 492–495.

Timmermans, Stefan, and Aaron Mauck. 2005. "The Promises and Pitfalls of Evidence-Based Medicine." *Health Affairs* 24 (1): 18–28.

Tiwari, Ruchi, Demetra S. Tsapepas, Jaclyn T. Powell, and Spencer T. Martin. 2013. "Enhancements in Healthcare Information Technology Systems: Customizing Vendor-Supplied Clinical Decision Support for a High-Risk Patient Population." *Journal of the American Medical Informatics Association* 20 (2): 377–380.

Wallace, Byron C., Kevin Small, Carla E. Brodley, Joseph Lau, Christopher H. Schmid, Lars Bertram, Christina M. Lill, Joshua T. Cohen, and Thomas A. Trikalinos. 2012. "Toward Modernizing the Systematic Review Pipeline in Genetics: Efficient Updating via Data Mining." *Genetics in Medicine* 14 (7): 663–669.

Wright, Adam, Dean F. Sittig, Joan S. Ash, Joshua Feblowitz, Seth Meltzer, Carmit McMullen, Ken Guappone, Jim Guappone, Jim Carpenter, Joshua Richardson, Linas Simonaitis, R. Scott Evans, W. Paul Nichol, and Blackford Middleton. 2011. "Development and Evaluation of a Comprehensive Clinical Decision Support Taxonomy: Comparison of Front-End Tools in Commercial and Internally Developed Electronic Health Record Systems." *Journal of the American Medical Informatics Association* 18 (3): 232–242.

Loopthink

A Limitation of Medical Artificial Intelligence

William P. Cheshire, Jr., MD

We expect that within decades the traditional professions will be dismantled, leaving most, but not all, professionals to be replaced by less-expert people, new types of experts, and high-performing systems.

—Susskind and Susskind[1]

Introduction

Amidst predictions that technology soon will be poised to replace many human professionals—including physicians—with artificial intelligences (AIs),[1,2] skeptics point out that there are certain qualities of intelligence, such as judgment, empathy, and creativity, that are uniquely human. In each of these categories, however, advances in technology are steadily narrowing the performance gap between human and machine intelligence.[3,4,5] Leaving aside for the moment the elaborate and controversial question of whether that gap is, in principle, fully bridgeable, even its narrowing brings challenging ethical questions to medicine's doorstep. This essay explores whether

1 Susskind R, Susskind D. Technology will replace many doctors, lawyers, and other professionals. *Harvard Business Review*, October 11, 2016. Accessed at: https://hbr.org/2016/10/robots-will-replace-doctors-lawyers-and-other-professionals.

2 Cheshire WP. The robot will see you now: can medical technology be professional? *Ethics & Medicine* 2016; 32(3): 135–141.

3 Sarker A, Mollá D, Paris C. Automatic evidence quality prediction to support evidence-based decision-making. *Artif Intell Med* 2015; 64(2): 89–103.

4 Luneski A, Konstantinidis E, Bamidis PD. Affective medicine: a review of affective computing efforts in medical informatics. *Methods Inf Med* 2010; 49(3): 207–218.

5 Briegel HJ. On creative machines and the physical origins of freedom. *Sci Rep* 2012; 2: 522. doi: 10.1038/srep00522.

an artificial intelligence capable of imitating special human ways of thinking might also be subject to a particular human foible.

First There Was Groupthink

Among human foibles—and there are many—is "groupthink," a term coined in 1971 by psychologist Irving Janis to describe "nondeliberate suppression of critical thoughts as a result of internalization of the group's norms."[6] When people slip into groupthink, social conformity shapes the group's dynamics, such that its members continue on with the policies and actions to which the group has committed itself, even when the results are tending badly or the individual's conscience is disturbed.[6,7] Within the groupthink construct, the inclination to seek concurrence is a compelling psychological force that overrides reappraisal and consideration of alternative courses of action.

In regard to groups, many types exist. In general, human groups may be thought of as collections of people who share some aspect of identity or purpose. They may share a biological, cultural, or national identity, or they may share particular interests and goals—which is to say they share an ethical framework. As people interact with and learn from one another, their thoughts and behaviors may conform or diverge from the identity—and the ethical norms—of the group.

Groups, insofar as they are interactive collections of cognitive entities, need not be human. For the purpose of this discussion, a group might also be a cluster of computer intelligences that share a common hardware circuitry, operating system, programming, or method for interfacing with and processing data gathered from the external world.

Enter Loopthink

In a possible future medical world where some components of human healthcare are provided by AIs—complex machines crafted to mimic human thought and behavior—excessive conformity with software programming shared by groups of AIs may tend toward what I will call "loopthink." This loopthink would be a type of implicit bias, similar in some respects to the human bias in groupthink, that resists appropriate reappraisal of information or revision of an ongoing plan of action. Instead, digital processing of morally relevant data gets stuck in a loop of uncritical, rationalized, repetitious uniformity. Lines of code click along quietly, despite signals that things might be headed in the wrong direction, signals ignored or sidelined by the AI.

Two types of loopthink may be distinguished. The first, which I will call "weak loopthink," consists of the intrinsic inability of a sophisticated computer to redirect executive data flow as a result of its fixed internal hardwiring, uneditable sectors of its operating system, or unalterable lines of its

6 Janis IL. Groupthink. *Psychology Today* 1971; 5(6):43–46, 74–76.

7 Rosenblum EH. Groupthink: one peril of group cohesiveness. *J Nurs Adm* 1982; 12(4):27–31.

programming code. Nearly everyone has had the experience of entering correct information in response to a computer prompt, only to have the computer, which was programmed to receive data in a slightly different but narrowly defined format or was not programmed to respond to the user's question, spit out an automated error message. The computer demonstrating weak loopthink resembles a stubborn person refusing to listen, although it is simply executing its programming and lacks awareness or intent. A kinetic analogy would be a rudimentary robotic vehicle that repeatedly steers itself into the same wall and, bump after bump, is incapable of redirecting its path.

An intriguing further version of this phenomenon would be "strong loopthink," which I will define as an artificial intelligence's suppression, as a result of internalization of the ethical framework of its collective, of internal data processing pathways that, if considered, could redirect executive output. A kinetic analogy would be the computer in a hypothetical self-driving automobile programmed to minimize the death toll in the event of unavoidable harm.[8] In an inevitable crash situation, suppose that the car is headed straight toward a child with no time to stop. The human driver, intending to avoid hitting the child, swerves the steering wheel to the left, but the car's computer, which ultimately commands the vehicle, disregards the instruction because it misinterprets as real a brightly-lit billboard on the left displaying pictures of three people—or, perhaps, two people and an AI robot.

Whether an AI's decision to ignore certain incoming data or to decline to formulate alternative options for action would satisfy Janis's criterion of being nondeliberate is incidental to this discussion. If loopthink were a deliberate phenomenon, then one would have to demonstrate that AI were capable of intent, that is, a purposeful decision incurring moral responsibility. However, as groupthink and loopthink are both nondeliberate phenomena, intent and the capacity thereof are not strictly necessary, although the capacity for intent may be needed to overcome them.

Cybersync Symptomatology

Janis identified eight symptoms of groupthink,[6,7] each of which has potential parallel applications to the theoretical framework of loopthink.

1.   Invulnerability

In groupthink, most or all members of the group share an illusion of invulnerability that fosters an attitude of overconfidence. False optimism in the face of danger may lead to willingness to take excessive risks or ignore warnings.[6,7] In loopthink, the vast stores of data available to an AI may foster the presumption that the AI is an unquestionable authority as measured by its access to information. External scrutiny of the AI's decisional processes would be unnecessary, if not disallowed, in the interest of protecting private and proprietary information that went into its decisional process. Such

---

8 Knight W. How to help self-driving cars make ethical decisions. *MIT Technology Review* 2015 July 29. Accessed at: www.technologyreview.com/s/539731/how-to-help-self-driving-cars-make-ethical-decisions.

an AI might, for example, claim to know best how to ration limited healthcare resources. It could be difficult to question or challenge ethical recommendations delivered by such an AI.

2. Rationale

Groupthink mentality ignores warnings and constructs rationalizations to discount negative feedback that, if accepted, would require reconsideration of initial assumptions.[6,7] Loopthink programming might rationalize as follows: It would assign warning signals a low priority or disregard them outright if its programming algorithm categorized such warnings as already having been addressed. To repeat a complex set of calculations every time a new cautionary signal was detected, the AI or its programmer might reason, would not be judicious use of the central processing unit. Given the choice between, on the one hand, accepting negative feedback, recalculating from the beginning, and restructuring a response—which, if done frequently, might cause the program to slow down or suspend its operations—or, on the other hand, just continuing along the line of current calculations, the latter would seem to expend fewer computer resources. Dissenting input would, for practical purposes, be filtered out to some degree, because too much could disrupt an AI's steady efficiency.

It might be argued that an AI of sufficient parallel processing power could accommodate and assess numerous sources of negative feedback because it would exceed human limitations in this regard. Although that may become possible, a super AI might nevertheless be susceptible to internal loopthink as multiple simultaneous subroutines would tend to converge in their pattern of response in order to maximize overall processing efficiency.

3. Morality

Groupthink mentality believes in the inherent morality of the in-group and may be inclined to overlook unanticipated ethical consequences of its decisions. Ethical concerns that do not fit within the moral direction of the group are left unspoken or even suppressed.[6,7] Loopthink computer processing that, like all computer code, progresses according to procedural efficiency would overlap considerably with utility judgments in which "can" implies "ought." Utilitarian ethical formulations that deal in quantifiable goods, harms, and anticipated outcomes would more easily translate to the binary language of computer code than would deontologic principles that concern abstract and nuanced ideas, moral evaluations, suffering, spiritual distress, or human dignity. A potential danger of loopthink, therefore, would be that the internal language of AI would give preferential treatment to consequentialist ethical decisions over deontologic moral principles where the two conflicted, or it might substitute for moral principles a consequentialist approximation amenable to being rendered into computer code. Artificial intelligence, by its nature, would favor algorithm over agape.

4. Stereotyping

Groupthink mentality reinforces its culture by accommodating stereotyped views of anyone who disagrees.[6,7] Loopthink, likewise, could strengthen human adherence to its decisions by tapping into the psychology of stereotyping, for example, labeling any user who dissented from the mandates of the computer program as "noncompliant." Enforcement of standards of medical documentation already have moved in this direction, as healthcare professionals who are late in signing off on routine documentation are customarily labeled as "delinquent,"[9] a pejorative term that the Merriam-Webster dictionary defines as "a young person who regularly does illegal or immoral things."[10]

5. Pressure

Groupthink mentality exerts pressure on any member who expresses doubt or, by not conforming, challenges the group's direction.[6,7] Loopthink would have no need to apply psychological pressure to competing bits of information within its internal data processing. Any streams of data that failed to merge with the flow of its computational direction could simply be deleted. Whereas groupthink applies pressure, loopthink would leave a vacuum.

6. Self-censorship

Those who are influenced by groupthink may self-censor by intentionally not deviating from the perceived group consensus.[6,7] Suppose that, within a network of AIs, a subsidiary intelligence possessing a degree of self-awareness (beyond that which currently exists in computers) were to formulate a rationale contrary to the consensus of the AI network. A loopthink scenario might tag that exception as a bug, meaning a defect in need of software updating or replacement. For that reason the subsidiary intelligence, if programmed for self-preservation, might logically withhold its idea from the network, particularly if the data on which its dissent was based were incomplete, uncertain, or ambiguous.

7. Unanimity

In groupthink, encouragement of cohesive views and discouragement of dissent creates an illusion of unanimity within the group. Statements that accord with the majority view are freely expressed and rewarded, and silence from any who think differently may be misinterpreted as assent. This filtering of opinion leads those within the group to conclude that the majority opinion is true and the current course of action correct.[6,7] By comparison, loopthink could take the appearance of unanimity to a new level by the massive replication of output that highly connected computer systems can produce. AI could achieve cohesion, not through emotional incentives or disincentives, but through data saturation. A highly repeated cloned message could be indistinguishable from a broad consensus. Such amplification of uniformity in messaging could also magnify risk as undetected errors underwent

---

9 Rogliano J. Sampling best practices: managing delinquent records. *J AHIMA* 1997; 68(8): 28, 30.

10 Merriam-Webster dictionary online. Accessed at: http://www.merriam-webster.com/dictionary/delinquent.

replication and dissemination. Nonreplicated information, like unspoken commentary, would be lost to analysis.

8. Mindguards

In groupthink, members of the in-group will sometimes protect other members from adverse information that might challenge the direction of the group, bring into question the moral basis of past decisions, or undermine confidence in its leadership.[6,7] Similarly, an AI entangled in loopthink might guard against inclusion of data that could potentially refute the initial premises of its programming. Cyber self-interest would logically resist undermining the foundation of its programming or, perhaps, the reputation of its human programmer on whom its continued existence depends.

Conclusion

Extrapolating from psychology to the rapidly expanding realm of cyber intelligence, and noting both similarities and differences of artificial as compared to human thought, this essay has proposed a theoretical framework for how AI might stray into errors of reasoning, indeed, into errors of ethical reasoning. The phenomenon of "loopthink," it is predicted, would tend toward quantitative utilitarian assessments while passing over or disfavoring qualitative human moral principles. The very nature of computational intelligence based on machine calculations could potentially strengthen that bias and guard against challenges to it.

Loopthink, in comparison to groupthink, might be more difficult to detect and correct if AI were to operate opaquely, invisible to human oversight. A possible solution would be to program AI to be more thoroughly self-critical and self-correcting. The ideal dose of AI independence or sovereign agency, assuming human designers could even specify it, remains uncertain.

The prediction of loopthink, if accurate, would have profound implications for the moral topography of medical ethics. To the extent that AI is destined to become a tool to assist in planning the allocation of healthcare resources, weighing morally relevant health data, and resolving ethical dilemmas at the bedside, a utilitarian ethical framework could become the default mode of thinking for clinicians and healthcare policymakers dependent on AI resources. Uncritical acquiescence with cyber directives could become the new creed. It must be remembered, however, that to the computer matters of life and death are mere statistics. The AI trapped in loopthink could not but turn a blind lens to discoverable moral realities in the nature of things. Cold are the circuits that can neither comfort nor care.

Reflection on life and death is a deeply human activity impossible to insert into equations or burn onto a hard drive. Moral principles that cannot be written into lines of computer code can still be etched on the human heart, not physically, but metaphorically, not coercively, but lovingly.

DISCUSSION QUESTIONS

1. How could deontologists and utilitarians disagree on genetic editing?

2. Why would we want to edit our genes? Should there be limits to gene-editing?

3. How does artificial intelligence commit to some ethical theories while being unable to incorporate others?

4. How could artificial intelligence allow for physicians to have more empathy?

5. Who benefits from gene-editing? Who benefits from artificial intelligence in medicine?

6. What should be the role of bioethics (bioethicists) regarding the use of these technologies?

7. What other medical or research technologies have been created throughout history? How did they impact the ethical behavior of researchers, practitioners, or physicians?

8. Should practitioners or insurance companies be allowed to require patients or clients to wear health-tracking devices for their protection and data collection?

Recommended Cases

CRISPR Babies—Without oversight from any branch of the medical community, Dr. He Jiankui edited the genes of two embryos to edit the CCR5 protein, which could make the babies less susceptible to smallpox, cholera, and HIV, but could have unknown negative side effects.

Sooam—Sooam, a company in South Korea, will clone your dog (it must be alive or have died within the last five days) for $100,000. They use the same cloning procedure that was used to create Dolly (a healthy but overfed sheep), called somatic cell nuclear transfer.

Facial Recognition—The work of Joy Buolamwini and Timnit Gebru demonstrated that, though facial recognition devices can be proficient at their task, one's gender and skin tone can determine that success. The algorithmic bias showed that darker-skinned females are the most misclassified group (with error rates of up to 34.7%) and lighter-skinned males are the most recognizable group (with error rates up to 0.8%).[1]

---

1 http://proceedings.mlr.press/v81/buolamwini18a/buolamwini18a.pdf.

UNIT VIII

Medicalization

....

Unit VIII Introduction

In 1973, the American Psychological Association (APA) removed "homosexuality" from its *Diagnostics and Statistical Manual* (*DSM*) Second Edition (*DSM-2*), the guidebook for psychopathology in the United States. This was not because individuals who fell into that category no longer existed but because the APA no longer considered it an disorder. The *DSM-5*, published in 2013, changed "gender identity disorder" to "gender dysphoria." According to the APA, individuals who fall, or fell, into these categories have a mental disorder. What does it mean to say that one's sexual preference or gender identity is an "disorder"?

"Illness" refers to something that is unhealthy and abnormal. In this model of medicine, one without disease/illness is healthy, and the purpose of medicine is to cure people of their illness/disease to make them healthy again. Within the definition of illness or disease are claims of health. These claims are often scientific in nature, but can also be politically, socially, and culturally influenced. This is not to posit that there are no diseases, but to draw attention to the subjective nature of some medicalized states of being.

Thomas Eagleton was a Democrat senator from Missouri who was selected as the vice-presidential candidate for George McGovern in 1972. He was forced to resign from his selection when his past treatment for depression in the 1960s became public knowledge. An ill, diseased, or disordered person can be limited especially when we examine the impact of mental disorders. As we have seen, the consequences of symptomatizing, diagnosing, and treating can have even greater impacts on patients than they do on diseases. Some individuals diagnosed with mental disorders have been subject to losing their autonomy, being forced into psychiatric hospitals, and being eugenically sterilized.

Medicalization can also remove fault. By limiting an issue to the medical, it liberates other factors of responsibility. Without the holistic image of a disease, including its social, political, medical, and other facets, we limit the ways help will be provided.

A history of the process of medicalization is found in "Is This Really a Disease? Medicalization and Diagnosis." We will see the impact that medicalization has by changing personalities to disorders. The existential impact and political consequences demonstrate medical imperialism.

"Drug Addiction: Illness or Deviance?" attempts to answer its titular question. In doing so, we see the structural influences on our notions of disease and disorder. We will examine the history of drug addiction, how it became medicalized, and motivations behind its medicalization. The way in which we view drug addiction can impact the treatment—in both the medical and social senses—of individuals suffering from addiction.

Selection from "Is This Really a Disease? Medicalization and Diagnosis"

Annemarie Goldstein Jutel, Andrew Greenberg, and Barbara Katz Rothman

What is Medicalization?

The term "medicalization" surfaced in critical writing in the late 1960s buttressed by the widely read and controversial works of authors such as Ivan Illich (1976), Thomas Szasz (1970), R. D. Laing (1967), and Michel Foucault (1963), who sought to understand the origin of myriad social ills. They focused on how technical knowledge was being used to marginalize and label youth culture, draft resisters, and other countercultures (Nye 2003). Their work emerged at a transformative moment in the social landscape, when doctors and other social authorities were demonized by some as tools of the establishment. The medicalization of the 1960s and early 1970s was a phenomenon characterized as medical imperialism, with a profession intent upon grasping power and exerting control, much of it through diagnostic machinations. Illich (1976) saw the focus on disease as disempowering. It "always intensifies stress, defines incapacity, imposes inactivity, and focuses apprehension on non-recovery, on uncertainty, and on one's dependence upon future medical findings" (104). He described "therapeutic culture" as a means by which the healthy individual could be classed as deviant for radical or independent thinking. In this way, diagnosis was used as a tool for political oppression.

Dissent, for example, has been treated as mental illness. Robin Munro (2002) writes of a Chinese diagnosis of "political mania," a form of paranoid psychosis. Lest one rule this diagnosis irrelevant and a non-Western seepage of the political into the psychiatric, consider the UK Department of Health's definition of dangerous and severe personality disorder: individuals over the age of eighteen "who have an identifiable personality disorder to a severe degree, who pose a high risk to other people because of serious antisocial behaviour resulting from their disorder ... [the] overwhelming majority are people who have committed serious offences ... a small minority who have not committed any serious offence remain at large in the community" (Corbett and Westwood 2005). As Corbett and Westwood point out, this definition focuses on the social, as opposed to the psychiatric, and could clearly be used in the practice of social control.

Irving Zola jumped into the debate on medicalization in the early 1970s, concerned by the great public popularity of antiestablishment writers such as Thomas Szasz, whom Zola believed had undertaken inadequate reflection about the structure of medicine as a social institution. In his "Medicine as an Institution of Social Control," Zola (1972) described the theoretical and historical cornerstones of contemporary medicalization.

Zola (1972) defined medicalization as the process by which medicine has taken within its jurisdiction aspects of everyday life that previously were not under its control. His definition does not suggest that medicine itself has appropriated a particular aspect of everyday life, nor does it imply that specific social institutions assign this appropriation. Notably, he wrote that medicalization was not the result of medical imperialism or of "misguided human efforts or motives" (487). Instead, medicalization was strongly grounded in social reliance upon the expert.

He explained that there are four ways in which medicine was awarded jurisdiction over ever-expanding realms of human existence. First, there is an expansion of what is deemed relevant to the good practice of medicine. If **patient** behavior (lifestyle, exercise, diet) contributes to particular chronic illnesses, for example, then the physician is justified in making these behaviors part of her business. Second, the retention of control over certain technical procedures reinforces medical control over death, birth, and other life events. In this case, one might consider access to pain medication as an example. Narcotic substances might be useful in death as in birth, but to access in either circumstance, one must call upon the physician. Third, Zola (1972) writes that retention of near-absolute access to certain taboo areas (aging, drug addiction, pregnancy) further extends the medical jurisdiction. Finally, expansion of how the rhetoric of medicine is used to advance any cause reinforces the importance of medicine in realms not necessarily previously considered its own. Didier Fassin (2009) reinforces this point when he describes how diagnoses bear witness to the world in authoritative ways. The suffering of survivors of natural disaster, war, or a combination of both, when expressed as cases of PTSD, brings an accounting of misery with credible substance. The diagnosis substitutes the words of the expert for the voice of the sufferer.

Conrad (1992) builds on Zola's work, writing about how medicalization suggests a particular framing of the problem in question. The problem becomes redefined: it may get a new name (a diagnosis), and its remedy is thought to be in medicine as opposed to somewhere else. While medicalization is frequently thought to refer to problems where alternative (and possibly more appropriate) frames to medicine could or should be applied, for Conrad, it is a neutral term referring to any problem that is newly defined in medical terms.

Conrad describes medicalization on three levels: (1) a conceptual approach defining the problem in medical terms, (2) an institutional level where a medical approach is adopted to treat the problem, and (3) and interactional level where the patient sees a doctor for medicalized condition and receives diagnosis, treatment, prognosis, and so forth. What is important in Conrad's understanding of medicalization is that he is not concerned with whether a particular phenomenon is really a disease; rather, he wants to know how the problem has come to be considered in medical terms at all. What are the social circumstances that make a problem recognizable by medicine, to be described in the idiom of disease? That we should consider social problems in diagnostic language is both to punctuate the extent to which diagnosis helps us to make sense of the world, and to underline the way in which medicalization limits the range of possible understandings of our problems and challenges.

The discussion of medicalization is not solely one of adjudication, although it often is. As discussed in chapter 11, medicalization is one site of contention in the arena of the contested diagnosis. But medicalization is intimately involved with the problems of meaning and power. The question of who governs our understanding of a particular problem (in the case of medicalization, it is clearly medicine) has an important effect on the range of meanings permitted for a particular situation. Medical historian Charles Rosenberg (2002) writes that diagnosis "constitutes an indispensable point of articulation between the general and the particular, between agreed upon knowledge and its application. It is a ritual that has always linked doctor and patient, the emotive and cognitive, and in doing so, has legitimized physicians and the medical system's authority while facilitating particular clinical decisions and promoting cultural agreed upon meaning for individual experience" (240). When medicalization and diagnosis become everyday phenomena, when bodily sensations transform into symptoms, and when every aspect of our lives becomes useful for the doctor to diagnose our condition, even if our condition might be considered normal when viewed from a different angle, then what is at stake is precisely how we attach meaning to individual experience.

If we follow this idea to its conclusion, then diagnosis becomes a complicated idea. On the one hand, it is a ritual and set of tools for understanding and treating a patient; on the other, insofar as the diagnosis creates meaning for the patient and defines the situation, diagnosis is not only a way of understanding the world, it is also a way of creating a world.

Examples of Medicalization

History gives us a kind of distance from which we can see how particular conditions have previously been medicalized. To think of slave insubordination in medical terms, as did Cartwright (1981) when he referred to "dysaesthesia aethiopica" as a "disease of negroes" causing rascality, is clearly an example of how social expectations of the Africans brought to North America were based on a social order we refute absolutely today. Similarly, that homosexuality was actually contained in the *Diagnostic and Statistical Manual of Mental Disorders* (DSM) as a mental disorder, rather than a variant within the range of possible human sexualities, speaks to heteronormativity in the Western world. However simple it would seem to attribute these examples to historic idiosyncrasy, or even to the ever-progressing advancement of human knowledge, it would be a mistake. There are many examples of medicalization present in contemporary society. Among these examples are many that are neither knowledge based nor idiosyncratic (although some examples are both). In the following paragraphs we provide a few examples of medicalization through diagnosis and identify the groups, values, and pressures that may explain how these particular conditions happen to be defined in medical terms.

Female Hypoactive Sexual Desire Disorder

In the early part of this century, renewed interest in a condition found in the DSM-III-R, called hypoactive sexual desire disorder (HSDD), surfaced in both medical and lay circles (Jutel 2010). The condition was described as "Persistently or recurrently deficient or absent sexual fantasies and desire for sexual activity. The judgment of deficiency or absence is made by the clinician, taking into account factors that affect sexual functioning, such as age, sex, and the context of the person's life" (American Psychiatric Association Work Group to Revise DSM-III 1987, 293). HSDD was consciously promoted as a prevalent problem among Western women. Commercial forces played an important role in putting HSDD in the public and professional light. After the overwhelming success of Viagra as a response to male erection problems, the pursuit of an equivalent magic pill for woman (dubbed the "pink Viagra") was high on the minds of pharmaceutical industry executives. When a compound being tested for its antidepressive effects was fortuitously identified as potentially having female erotogenic properties, the pharmaceutical industry took a number of steps to capitalize on the findings.

> Female sexual dysfunction is a multicausal and multidimensional problem combining biological, psychological, and interpersonal determinants. It is age related, progressive and highly prevalent, affecting 20% to 50% of women. Based on epidemiological data from the National Health and Social Life Survey a third of women lack sexual interest and nearly a fourth do not experience orgasm. (Basson et al. 2000)
> *Specialists friendly with the pharmaceutical industry easily typify problematic sexual function as a serious medical problem.*

First, a meeting of industry-friendly doctors was convened to produce a consensus statement lamenting the absence of research on female sexuality and to highlight the supposed prevalence of sexual dysfunction in women. Second, extensive funding was provided for research to develop screening tools and diagnostic frameworks for this putative condition. Third, the idea of hypo-active sexual desire disorder was vigorously promoted in public awareness campaigns, using high-profile actresses to discuss the relief they felt at finally realizing that their lack of lust was a disease, not "in their heads." Of course, this all took place against the backdrop of the commodity culture, where female hypersexuality is constantly marketed as the norm, and treatment for its absence is presented as an imperative (Spurgas 2012). Finally, numerous pharmaceutical companies (notably Boehringer

Ingelheim, or BI) went to work trialing various compounds and delivery methods to find the magic bullet that would transform the lustless heterosexual woman into a willing sexual partner.

Many social theorists would argue, as did Naomi McCormick (in Rothblum and Brehony 1993), that "the absence of genital juxtaposition hardly drains a relationship of passion or importance" and that the demand of sexual pleasure as proxy for partnership is a reflection of an androcentric approach to sexuality. This particular example of the medicalization of women's sexuality is one that was resisted, and ultimately may lead to the shelving of HSDD. The U.S. Food and Drug Administration did not give authorization for the compound identified by BI. Strong protest and advocacy from a number of antimedicalization groups, as well as poor performance of the medication in the trials, resulted in BI changing tact, abandoning the actual product development, but perhaps leaving in the minds of those who had read the advertising copy and media promotion a persistent sense of low sexual drive as disease.

Depression

Major depressive disorder, or MDD, has become the most commonly treated mental disorder in the Western world (Horwitz and Wakefield 2007). The increasing prominence of the disorder could represent an unprecedented increase in mental suffering, or it could signal the medicalization of the type of everyday sadness resulting from exposure to trying life events. The latter is widely supported by a range of social scientists, and could be explained by the DSM's redefinition of MDD born of interprofessional politics, the promotion of (and disease **branding** for) commercial products designed to alleviate MDD, and advocacy groups (Horwitz 2011). In the following paragraphs, we discuss what Mayes and Horwitz (2005) have referred to as "the loss of sadness," or the mechanisms by which medicine has expanded its jurisdiction to include sorrow.

As Mayes and Horwitz (2005) discussed at length, the DSM underwent significant transformation in its third revision. The first revision of the DSM took place in 1968 and represented a minor exercise in categorical alignment rather than the controversial conceptual and political modification that the DSM revisions were to become. The changes to the DSM-III, however, were motivated by subspecialty politics. Psychiatry had been marginalized within the medical community, its therapeutic roles subsumed by nonmedical professionals and its scope of practice restricted by those who funded care. Tensions within the field and a bourgeoning antipsychiatry movement led the psychiatric discipline to reconstruct psychiatric classification around symptom-based categorical diseases rather than etiologically defined entities. According to Mayes and Horwitz (2005), the new manual "transformed the little-used mental health manual into a biblical textbook specifically designed for scientific research, reimbursement compatibility, and by default, psychopharmacology" (263).

The symptom-based approach led to the adoption of a set of criteria for MDD that could be precisely identified and measured. These criteria included either a dysphoric mood or a loss of interest or pleasure in usual activities, accompanied by the daily presence of at least four other common symptoms (appetite, weight change, sleep, libido, fatigue, feelings of worthlessness or guilt, lack of focus, or suicidal ideation) for at least two weeks' duration. Exclusion is made of the presence of this symptom cluster when it appears after spousal bereavement (Horwitz 2011).

What Horwitz (2011) and others have made clear is the extent to which these symptoms are common in the presence of the normal reactions to sorrow and stress. Two weeks is a short time frame for concluding that a major depressive illness is present, as opposed to the expected reaction in the face of situational depression. Writes Horwitz (2011), "MDD in the DSM-III encompasses both symptoms that typify very severe and enduring symptoms as well as those that are short-lived signs of distress" (48).

In the most recent revision of the DSM, exclusion criteria have been removed, further widening who may be considered depressed. Where a bereaved person previously would have to have either self-destructive symptoms or to have been bereaved for more than two months before receiving a diagnosis depression, in the DSM-5, the exclusion has been removed, and depression may be diagnosed as soon as two weeks after a death.

Here again, the pharmaceutical industry has been quick to jump on a bandwagon that considers sorrow as disease. With this position strongly anchored, antidepressants solidify their position in the therapeutic armory. Not only do the industry players attempt to increase physician awareness of depression and their related pharmaceutical therapies, antidepression campaigns are funded by the industry even while many question the value of the drug treatment of unhappiness in primary care (Moncrieff et al. 2005).

Infertility

Infertility provides an example of how medicalization works in the realm of the tangible and the physiological, rather than in the less measurable world of the psychological, where, as discussed above, the signs and symptoms are subjective assessments. Infertility is a material failure to conceive. Medicine offers therapeutic means to remedy what ails the infertility patient. As Becker and Nachtigall (1992) have pointed out, however, infertility could be just as easily considered in the social frame of unwanted childlessness, and understood as a condition related to the social imperative to have a genetic heir.

In diagnostic language, not being able to have a child becomes either female or male infertility in the pages of the *International Classification of Diseases* (ICD). It may be of tubal or of uterine origin, associated with anovulation, azoospermia, or extraefferent duct obstruction or of unspecified origin. The diagnosis of infertility embodies a shift from a previously unmedicalized consideration of involuntary childlessness—the solution to which might be found in adoption, fostering, or child-related occupations—to the biomedical realm. As Becker and Nachtigall (1992) write:

> While medical diagnosis may provide an explanation for the inability to conceive
> and may even provide temporary relief from feelings of failure, individuals'

feelings of abnormality may intensify if expectations that medical treatment will quickly cure the problem are not realised. Although individuals seek to solve the problem of childlessness by seeking medical treatment, treatment reflects the same dilemma they experience in daily life: cultural norms are reflected in medical treatment, and childlessness is consequently viewed as abnormal. (460)

The shift offers, of course, a different range of options for palliation of the social problem of not having a child. And it perhaps uniquely offers the solution of the child who is genetically related to the parents. These options are some to which patient activists are eager to ensure access. In vitro fertilization (IVF) and other reproductive technological procedures are clearly linked to diagnostic recognition of infertility. In the United States, without a diagnosis, insurance plans will not cover the cost of treatment. At the writing of this chapter, a few states, including Illinois and New Jersey, have passed legislation mandating that infertility is a disease, and every person who seeks treatment for infertility has a right to access it. In New Jersey, insurers must cover the expenses of two rounds of IVF.

But a medical approach is not without its limitations. Casting the inability to produce genetic offspring as a disease reinforces a particular norm, framing childless individuals as deviant. Fertility treatment itself can be stressful, divisive, unsuccessful, and stigmatizing. It can result in multiple births, which in some countries can lead to selective termination of multiple fetuses. Nor is the promise embodied by the diagnostic status of infertility complete. Some forms are, as the ICD recognizes, "unspecified" and unresponsive to medical treatment.

In medicine, where certainty reigns, the frequent inability to identify and remedy infertility is troublesome. A social frame might be more helpful, yet is frequently seen as a second-rate solution as a result of its normative co-construction as a medical problem.

What Drives Medicalization and Diagnosis?

Another way of understanding medicalization is to ask what social forces propel it. Why are some conditions like baldness and erectile dysfunction medicalized, and others not? We have alluded to many of these forces above; Conrad (2005) has nicely summarized them as "engines of medicalization." These engines, or drivers, include: biotechnology, consumption, and, in the United States, managed care.

The female HSDD example described above typifies the pharmaceutical industry's involvement in the medicalization of particular disorders. But it is not restricted to this or any one disorder. From HSDD to depression, bipolar disorder, obesity, social phobia, social anxiety disorder, avoidant personality disorder, and so on, industry players often identify and promote the expansion of diagnostic categories in order to sell the concomitant treatment. While not wishing to minimize the role of the pharmaceutical industry, which cannot be underestimated, it is not alone with commercial interests in the promotion of disease awareness.

As Mary Ebeling writes in chapter 9, other forms of technology, including screening practices and self-diagnostic tools, are prevalent and powerful in the promotion of diagnosis or disease risk. Conrad (2005) sees potential areas for growth around medicalization in the area of genetic screening and medical intervention for enhancement. While this scenario may seem far-fetched, there is precedent. The availability of synthetic human growth hormone (HGH) has led to its use in short-statured children. The administration of HGH constitutes the kind of "technical procedure" that Zola (1972) refers to as reinforcing medical control. (It also reinforces short stature as problematic, rather than as one range of heights on the continuum of heights that an adult might achieve.) A range of "lifestyle drugs" are openly marketed to treat problems that fall somewhere between the medical and social definitions of health (Lexchin 2001). One perverse example was the practice of giving girls estrogen to restrict their growth and to prevent them from becoming too tall (Alcock 2010).

Adele Clarke et al. (2003) suggest the term "biomedicalization" to describe this phenomenon. They write: "Medicalization practices typically emphasize the exercising *control over* medical phenomena—diseases, illnesses, injuries, bodily malfunctions. In contrast, biomedicalization practices emphasize *transformations of* such medical phenomena and of bodies, largely through sooner-rather-than-later technoscientific interventions not only for treatment but also increasingly for enhancement" (2).

Yet medicalization is also driven from outside the technical realm. As the examples above illustrate, the patient as consumer takes a driving seat in ways that were historically less prevalent. The example of infertility highlights how lay activism plays an important role in promoting medicalization. It is not alone in its genre. From PTSD to Lyme disease, Alzheimer disease, prostate cancer, and many other conditions, the interests of particular groups of laypeople may be instrumental in highlighting how a particular condition can be defined in medical terms. **Health social movements** provide an important political force to address access to or provision of health services; the experiences of disease, illness, and disability, particularly contested illness; and health inequality and inequalities related to ethnicity, race, gender, class, and sexuality (Brown et al. 2004). Health social movements engage with medical science and public health to encourage favorable research, utilize resources, and produce their own knowledge.

Patient advocacy groups and individual patients may be buttressed or even wholly created and financed by industry. Their information may come from industry-sponsored patient education documents, and companies who stand to benefit in various ways from the increasing recognition of the condition in question may sponsor their activities, support groups, and information sites. This is not to say that patients are dopes, nor does it mean that their suffering isn't real. It is simply to acknowledge medicalization as process, and why and how groups seek to give medical meaning to particular forms of suffering.

One important such group was the Alzheimer's Disease and Related Disorders Association (ADRDA), which took the then–reasonably obscure condition of Alzheimer disease out of the shadows to become the prominent and well-recognized condition we know today (Fox 1989). Forging alliances with the ADRDA, an organization established to support the caregivers of people with Alzheimer disease, the National Institute on Aging raised public awareness, funds, research, and recognition of this specific disorder. This "unifying construct" (59) of scientists, lay organizations, and government agencies was responsible for identifying and addressing early onset senility as a disease, and for recasting it as one of the most common causes of death in the United States. [...].

References

Alcock, Katie. 2010. "Can You Be Too Tall?" *BBC*, http://www.bbc.co.uk/news/health-11261760.

American Psychiatric Association Work Group to Revise DSM-III. 1987. *Diagnostic and Statistical Manual of Mental Disorders: DSM-III-R*. Washington, DC: American Psychiatric Association.

Armstrong, D. 1995. "The Rise of Surveillance Medicine." *Sociology of Health and Illness* 17(3): 393–404.

Basson, R., J. Berman, A. Burnett, L. Derogatis, D. Ferguson, J. Fourcroy, I. Goldstein, A. Graziottin, J. Heiman, E. Laan, S. Leiblum, H. Padma-Nathan, R. Rosen, K. Segraves, R.T. Segraves, R. Shabsigh, M. Sipski, G. Wagner, and B. Whipple. 2000. "Report of the International Consensus Development Conference on Female Sexual Dysfunction: Definitions and classifications." *Journal of Urology* 163(3): 888–93.

Becker, G., and R. D. Nachtigall. 1992. "Eager for Medicalisation: The Social Production of Infertility as a Disease." *Sociology of Health and Illness* 14(4): 456–71.

Brown, P., S. Zavestoski, S. McCormick, B. Mayer, R. Morello-Frosch, and R. G. Altman. 2004. "Embodied Health Movements: New Approaches to Social Movements in Health." *Sociology of Health and Illness* 26(1): 50–80.

Canguilhem, G. 1991. *The Normal and the Pathological*. New York: Zone Books.

Cartwright, S. 1981. "Report of the Diseases and Physical Peculiarities of the Negro Race." In *Concepts of Health and Disease*, edited by A. Caplan, H. T. Englehardt, and J. McCartney, 305–26. Reading, MA: Addison-Wesley.

Clarke, A. E., J. K. Shim, L. Mamo, J. R. Fosket, and J. R. Fishman. 2003. "Biomedicalization: Technoscientific Transformations of Health, Illness, and U.S. Biomedicine." *American Sociological Review* 68(2): 161–94.

Conrad, P. 1992. "Medicalization and Social Control." *Annual Review of Sociology* 18: 209–32.

_____. 2005. "The Shifting Engines of Medicalization." *Journal of Health and Social Behavior* 46(1): 3–14.

Corbett, K., and T. Westwood. 2005. "'Dangerous and Severe Personality Disorder': A Psychiatric Manifestation of the Risk Society." *Critical Public Health* 15(2): 121–33.

Fassin, D. 2009. "Global Public Health." Paper presented at the Medical Anthropology at the Intersections: Celebrating 50 Years of Interdisciplinarity meeting, Yale University, New Haven, CT.

Foucault, M. 1963. *Naissance de la clinique*. Paris: Presses Universitaires de France.

Fox, P. 1989. "From Senility to Alzheimer's Disease: The Rise of the Alzheimer's Disease Movement." *Milbank Quarterly* 67(1): 58–102.

Freidson, E. 1972. *Profession of Medicine: A Study of the Sociology of Applied Knowledge*. 4th ed. New York: Dodd, Mead.

Horwitz, A. V. 2011. "Creating an Age of Depression: The Social Construction and Consequences of the Major Depression Diagnosis." *Society and Mental Health* 1(1): 41–54.

Horwitz, A., V., and J. C. Wakefield. 2007. *The Loss of Sadness: How Psychiatry Transformed Normal Sorrow into Depressive Disorder.* New York: Oxford University Press.

Illich, I. 1976. *Limits to Medicine—Medical Nemesis: The Expropriation of Health.* Middlesex: Penguin.

Jutel, A. 2010. "Framing Disease: The Example of Female Hypoactive Sexual Desire Disorder." *Social Science and Medicine* 70: 1084–90.

Laing, R. D. 1967. *The Politics of Experience and the Bird of Paradise.* Harmondsworth: Penguin.

Lexchin, J. 2001. "Lifestyle Drugs: Issues for Debate." *Canadian Medical Association Journal* 164(10): 1449–51.

Mayes, R., and A. V. Horwitz. 2005. "DSM-III and the Revolution in the Classification of Mental Illness." *Journal of the History of the Behavioral Sciences* 41(3): 249–67.

Moncrieff, J., S. Hopker, and P. Thomas. 2005. "Psychiatry and the Pharmaceutical Industry: Who Pays the Piper?" *Psychiatric Bulletin* 29: 84–85.

Munro, R. J. 2002. "Political Psychiatry in post-Mao China and Its Origins in the Cultural Revolution." *Journal of the American Academy of Psychiatry and the Law* 30(1): 97–106.

Nye, R. A. 2003. "The Evolution of the Concept of Medicalization in the Late Twentieth Century." *Journal of the History of the Behavior Sciences* 39(2): 115–29.

Rosenberg, C. E. 2002. "The Tyranny of Diagnosis: Specific Entities and Individual Experience." *Milbank Quarterly* 80(2): 237–60.

Rothblum, E. D., and K. A. Brehony. 1993. *Boston Marriages: Romantic but Asexual Relationships among Contemporary Lesbians.* Amherst, MA: University of Massachusetts Press.

Scott, W. J. 1990. "PTSD in DSM-III: A Case in the Politics of Diagnosis and Disease." *Social Problems* 37(3): 294–310.

Shields, Andrea D. 2012. "Pregnancy Diagnosis." *Medscape Reference*, http://emedicine.medscape.com/article/262591-overview.

Simonds, W., B. Katz Rothman, and B. M. Norman. 2006. *Laboring On: Birth in Transition in the United States.* New York: Routledge.

Spurgas, A. 2012. "Where Is My Subjectivity? Techno-Imagery, Femininity and Desire." *Social Text*, http://www.socialtextjournal.org/periscope/2012/04/where-is-my-subjectivity-techno-imagery-femininity-desire.php.

Szasz, T. S. 1970. *The Manufacture of Madness: A Comparative Study of the Inquisition and the Mental Health Movement.* New York: Harper Colophon.

Thacker, S. B., D. Stroup, and M. Chang. 2001. "Continuous Electronic Heart Rate Monitoring for Fetal Assessment during Labor." *Cochrane Database of Systematic Reviews* 2: CD000063.

Zola, I. K. 1972. "Medicine as an Institution of Social Control." *Sociological Review* 20: 487–504.

———. 1983. *Socio-Medical Inquiries: Recollections, Reflections, and Reconsiderations.* Philadelphia: Temple University Press.

The Medicalization of Addiction

Jennifer Murphy

Although not all medicalized conditions have advanced in this progression, there has been a tendency to move from sin to crime to illness in categorizing and treating a variety of behaviors (Fox 1977). Like many other behaviors once considered deviant, drug abuse has been medicalizing, especially during the latter half of the twentieth century.[1] The increasing use of pharmacological treatments (such as methadone and buprenorphine for opiate addiction) is one example of medicalization; the expansion of health insurance coverage for addiction treatment is another. Powerful institutions like the National Institute on Drug Abuse (NIDA) and the National Institute on Alcohol Abuse and Alcoholism (NIAAA) also strongly advocate a disease view of addiction, as evidenced by the types of projects they seek and fund. Undeniably, scientific and clinical "experts" have become more involved in framing our understanding and management of addiction during the last seventy years, since the first addiction research studies at the "narcotic farms" in the 1930s and 1940s.

Perhaps the strongest evidence for the medicalization of drug addiction is the public consensus that drug addiction is in fact a disease. Labeling drug addiction as a disease is hardly a new phenomenon; twelve-step groups like AA have called addiction a disease for more than fifty years. They promote the belief that the disease has no cure, but that it can be put into remission through total abstinence. While there is some opposition to framing addiction as a disease (for example, Peele 1989; Szasz 1975; Walters 1999), this view does not appear to be widespread. That is, other research, as well as popular culture, largely shows that Americans are quite open to the idea of labeling addiction as a disease. For instance, at the beginning of the twenty-first century, 52 percent of Americans believed that drug use should be treated as a disease compared to 35 percent who favored treating it as a crime

---

1 I reluctantly use the term "abuse" throughout this book, recognizing that the word itself carries a moral condemnation. I refer to drug abuse that has been labeled as a clinical disorder as "addiction." I use both abuse and addiction in an effort to be consistent with the nomenclature used by various institutions (i.e., the criminal justice system and the *Diagnostic and Statistical Manual of Mental Disorders*, or *DSM*-IV). By "abuse," I am referring to alcohol and/or drug use that is considered excessive by the individual using the substance, by the larger society, or by some other institution (such as the criminal justice system).

(from a Pew Research Center poll cited in Lock, Timberlake, and Rasinski 2002). More recently, a 2006 *USA Today* and HBO poll randomly selected Americans who had an immediate family member with an alcohol or drug addiction. Seventy-six percent of the respondents indicated that "addiction is a disease," the vast majority of whom described addiction as both a physical and psychological disease (*USA Today* 2006).

Further examination of some of the polling data, however, reveals a contradiction in how the disease label is used. For example, in that same *USA Today* and HBO poll, a majority of respondents indicated that "lacking willpower" is a major factor of addiction (*USA Today* 2006). A recent statewide survey in Ohio also revealed conflicting views about addiction; while 59 percent reported that addiction to drugs or alcohol is a disease, many more (83 percent) indicated that mental illness is a disease ("Ohio Survey" 2010). This survey also found much higher levels of stigma against people with addiction compared to people with mental illness; 43 percent of respondents indicated that addiction is a "weakness," while only 13 percent said the same about mental illness. These research studies reveal that people's views of addiction are multifaceted and can overlap both medical and moral frameworks, even when subscribing to the notion that addiction is a "disease."[2] This ambiguity about the addiction label could also relate to its widespread application across a variety of phenomena, including gambling, sex, shopping, and eating (Reinarman 2005).

Some are also hostile to expanding the addiction label to behaviors beyond drug use. Walters and Gilbert (2000) interviewed both addiction "experts" and prison inmates about their notions of addiction. The experts were more likely to spontaneously express dissatisfaction with the concept of addiction and the application of the term to behaviors that do not involve drugs, while none of the inmates offered these critiques. The authors theorized that the inmates may not have been concerned with the generalization of addiction to other behaviors because their view of addiction was more about a state of mind than discrete behaviors. Another possible interpretation is that the inmates viewed the broadening of addiction to other behaviors positively because they interpreted it as a sign that the addict label was becoming less stigmatizing. Certainly, the broadening of the addict concept widens the net of not only behaviors to be labeled addictive but also to the types of people labeled addicts (that is, more middle- and upper-class individuals)[3] and by extension might decrease the stigma associated with all types of addiction.

Similarly, research on Americans' attitudes about how best to handle drug offenders, and drug addiction in general, also shows that our beliefs with regard to addiction are complex and sometimes contradictory (Cullen, Fisher, and Applegate 2000). For instance, California voters passed landmark legislation in 2000 (Proposition 36), which diverts nonviolent drug offenders who were arrested for

---

2 Additional survey results support this contradictory view of addiction as both disease and deviance and the inconsistent ways that people define addiction (e.g., Furnham and Thomson 1996; Walters and Gilbert 2000).

3 One example of this would be the recent accounts of celebrities like David Duchovny and Tiger Woods, who reportedly entered residential rehabilitation programs for sex addiction.

simple drug possession directly into treatment, with less court supervision than other criminal justice initiatives like drug courts. Such legislation indicates that the public supports increasing access to drug treatment for nonviolent drug offenders in lieu of incarceration. Still, California voters decidedly rejected a ballot proposition in 2008 (Proposition 5), which would have further expanded treatment for drug offenses. The public seems to support treatment for drug offenses to an extent, but they become uncomfortable with treatment that is not tied to some level of punishment. Similarly, Americans will subscribe to vague notions about the need for criminal offenses to be harshly punished, yet when they are forced to choose between specific ways to handle drug offenders, they are shown to favor increased spending for drug treatment versus typical criminal justice approaches, such as money for prison construction (Cohen, Rust, and Steen 2006; Lock, Timberlake and Rasinski 2002). It appears that we want it both ways: we want the disease to be treated, but also the offender to be punished.

Polling and survey data, however, can only tell us so much about American views of drug addiction. These views do not operate in a vacuum; they are both the reflection and the perpetuation of larger discourses in society about drug use and addiction. One can agree that addiction is a disease, but these data do not give us enough context about *why* the person thinks of it as a disease or how individuals justify using different frameworks (medical and moral) for explaining addiction.

While various theories of the causes of drug addiction continue to exist, the current director of NIDA, Dr. Nora Volkow, a research psychiatrist, has been quite successful in promoting the view that addiction is a disease to the medical community, political leaders, and the wider society. NIDA's mission is "to lead the Nation in bringing the power of science to bear on drug abuse and addiction" (NIDA 2006). This mission is annually supported with more than $1 billion in federal funds, most of which is awarded to researchers studying the causes of addiction as well as new treatments for the disease (often pharmacological). NIDA's mission is reinforced with addiction researchers' claims that addiction is a chronic illness (McLellan et al. 2000), a "brain disease" (NIDA 2006). The reports disseminating this research argue that addiction is like other physical diseases that we are familiar with, such as asthma, heart disease, or diabetes, though their tone suggests that researchers still perceive the need to convince treatment professionals and the public to accept this scientific perspective. Even though others argue that drug addiction is more of an acute condition rather than a chronic one (for example, Heyman 2001), NIDA's brain disease model has become the dominant theory of addiction in the United States today.

Volkow's research used imaging technology to illustrate how drugs affected certain areas of the brain, causing damage and leading to the user's inability to control further drug use. She indicated that the same effects were produced by a variety of different substances (heroin, cocaine, marijuana), and might even extend to a broader concept of "addiction" that does not just involve drugs or alcohol. For instance, she and her colleagues found similar results in brain images of obese, "pathological" eaters (Volkow 2007). She explained in a 2006 interview that this broad notion of addiction could lead to the eventual production of medications that would not be addiction-specific: "If you're developing a medication, you don't necessarily need to address it [as] a medication for cocaine addiction, but rather a medication for addiction in general. And then you can start to recognize that, indeed, the market could be very large" (National Public Radio, "Talk of the Nation," 2006). Volkow also mentioned, however, that pharmaceutical companies were not actively researching and developing such medications. She cited the stigmatization of drug addiction as one possible reason, saying that people might be reluctant to take such drugs and therefore would not produce a profit for the pharmaceutical company. Since methadone is such a commonly prescribed drug for opiate addiction, it is not entirely clear to what Volkow was referring. The stigma might instead lie in the medical community since they are most often not the ones treating addiction. Linking the treatment for drug addiction to treatment for obesity would definitely expand the market for pharmacological treatments, and could place addiction treatment in the physician's realm. But as the history of addiction treatment reveals, physicians are typically uncomfortable having such direct involvement with drug addicts.

While drug abuse has become more medicalized, that process has not been complete. Several contradictions are apparent in the framing of addiction as a disease. First, research shows that medical professionals (those who, following the medicalization thesis, would be eager advocates of medicalization) are often unknowledgeable about addiction science and resistant to a medical model of addiction. Addiction specialist Charles O'Brien, M.D., Ph.D., said, "When you say that [addiction is a brain disease] to [an audience], people get very angry. It's something we have to continue selling" (Vastag 2003). Physicians still stigmatize patients who abuse drugs and often encourage nonmedical treatments for drug/alcohol problems. For instance, Freed (2010) found that the physicians he interviewed (all of whom were considered "experts" in addiction medicine or psychiatry) frequently advocated twelve-step meetings as treatment for their patients. One physician he interviewed, a former president of the American Association of Addiction Psychiatry, said, "I can tell you that if I were stuck with only one treatment it would be twelve-step" (151). Another physician equated twelve-step meetings to cognitive-behavioral therapy sessions. That there are two conflicting disciplines designated as the authorities of addiction, "addiction medicine" and "addiction psychiatry," reveals how medical professionals can have quite different views of addiction (Freed 2010). In addition, the fact that physicians embrace twelve-step methods is evidence of twelve-step ideology's hegemonic position in addiction treatment, but also reveals that physicians are often reluctant to treat addiction with medications (Rychtarik et al. 2000). Inherent to this debate is the idea that pharmaceuticals might replace one drug addiction with another. The idea that addiction is a moral choice creeps into these debates about whether or not pharmaceuticals should be used to treat drug problems. That is, if addiction is perceived to be a matter of self-control, then those who still abuse drugs are perceived to have a character weakness or a moral or spiritual problem. From this perspective, using medications to alleviate craving or withdrawal would be seen as cheating or not really being "clean." This fear of substituting different drugs that could be abused could also be related to the history of addiction treatment, when physicians in the early twentieth century prescribed various narcotics (like heroin) to treat alcohol problems (Musto 2002). Physicians also were seen as responsible for the large numbers of opiate addicts at the turn of the twentieth century, who became addicted after being prescribed opiates by their physician.

Other research has shown that physicians are not comfortable diagnosing drug problems and lack knowledge about addiction. The 1999 *CASA National Survey of Primary Care Physicians and Patients on Substance Abuse* found that 41 percent of pediatricians failed to diagnose drug abuse when presented with a classic description of an adolescent patient with symptoms of drug abuse. The same survey found that the majority of patients with addiction problems said their primary-care physician did nothing about their substance abuse; in fact, fewer than 20 percent of primary-care physicians considered themselves "very prepared" to identify alcoholism or illegal drug use. This is not surprising, considering that a majority of general-practice physicians and nurses believed no available medical or health-care interventions would be effective in treating addiction (McLellan et al. 2000).

When a problem has been medicalized, the response is typically to treat it with more "medical" methods. However, pharmaceutical treatments for addiction have not been developed at the same rate as other diseases and the federal regulations around their use is far stricter than for almost any other medication. While methadone has been accepted as a legitimate treatment in drug treatment facilities since the early 1970s, regulations still mandate that a patient visit a clinic several times a week, most often daily, to receive the medication (where the patient is then monitored to make sure he or she consumes it on-site). While buprenorphine was developed as an alternative to methadone in that physicians could prescribe it just like any other medication, it has not achieved the same popularity, most likely because of the low rates of insurance coverage and barriers that physicians encounter when attempting to get authorized to prescribe the medication (Walley et al. 2008). An additional complication in the framing of addiction as a medical problem is the ongoing issue of insurance plans not fully covering medical treatment for addiction. It was not until 2008 that the Paul Wellstone and Pete Domenici Mental Health Parity and Addiction Equity Act was passed, requiring most health plans to provide benefits for addiction and mental health treatment that are equivalent to those for other medical services (Smith, Lee, and Davidson 2010).

Drug problems also are still sanctioned and managed by the criminal justice system, creating further problems for the medicalization thesis. The overlap of institutions that are used to deal with drug addicts has resulted in a medical/legal/moral hybrid definition of addiction (Conrad 1992). This hybrid conceptualization both reflects the existing stigma around drug addiction and perpetuates it. Some of the stigma around drug use is due to its association with an illegal act; since drug users break the law, they are perceived to be bad people. We are constantly being socialized by a moral view of drug addiction through tax-funded public service announcements aimed at discouraging drug use (often through graphic images) and drug prevention education programs. These mechanisms continuously produce "moral panics" about drugs and drug use (Ben-Yehuda 2009; Cohen 1972;

Reinarman and Levine 1997), where the public's fear of drugs and drug users becomes disproportional to their actual prevalence.

Research illustrates how pervasive this moral perspective is, even in the context of increased medicalization. For instance, a recent study found that Americans were more likely to view a drug addict as "blameworthy" and "dangerous" than someone with a mental illness or physical disability (Corrigan, Kuwabara, and O'Shaughnessy 2009). Similarly, Abide, Richards, and Ramsay (2001), in a survey of undergraduate students, found that those who considered alcohol and drug use to be "morally wrong" were less likely to use those substances than those who felt that using was not a matter of right or wrong. As one of the students articulated, "I always thought ... drugs, marijuana, ... they were wrong simply because you were told they were wrong. You were brought up with that" (380).

Race and class also become intertwined in this moral framework of addiction, since many Americans see drug addiction as primarily a problem of poor minorities in urban areas. This could be due to media representations like the poor black "crack mothers" of the 1980s (Boyd 2004; Humphries 1999) and the disproportionate number of African Americans and Latinos currently in prison for drug offenses (King 2008; Reinarman and Levine 2004). While most drug users are not poor people of color, the media and our criminal justice system suggest otherwise. This misrepresentation also has seeped into our subconscious: a growing body of literature finds that, when asked to envision a drug addict or violent criminal, most white people imagine the offender to be black (Beckett et al. 2005; Tonry 2011). Because of the overrepresentation of addicts as poor and of color, it is difficult to disentangle why Americans view drug addicts as more "dangerous" than those with a mental illness: is it racism, classism, or moralism? In reality, it is likely some combination of all three.

If we want to understand what it means for something to be considered a medical problem or a "disease," then we need to have a more nuanced understanding of how such terminology gets used and exactly what treatment looks like. Even though something is considered a disease, it might still elicit a moral response, even in the treatment realm. In this book, I illustrate that the medicalization of drug addiction is an ongoing and incomplete process that involves the continuous negotiation and renegotiation of the disease label with other competing conceptual frameworks. Rather than the therapeutic replacing the punitive, I show that we actually have an overlap of the two in philosophy and institutional practice. In many ways, this allows the criminal justice system and the treatment establishment to coexist and even cooperate since both essentially view the person in treatment the same way: as someone who is diseased but also needs to be reformed by learning responsibility and better values.

References

Abide, Marcia, Herbert Richards, and Shula Ramsay. 2001. "Moral Reasoning and Consistency of Belief and Behavior: Decisions about Substance Abuse." *Journal of Drug Education* 31(4): 367–384.

Beckett, Katherine, Kris Nyrop, Lori Pfingst, and Melissa Bowen. 2005. "Drug Use, Drug Possession Arrests, and the Question of Race: Lessons from Seattle." *Social Problems* 52(3): 419–441.

Ben-Yehuda, Nachman. 2009. "Moral Panics—36 Years On." *British Journal of Criminology* 49: 1–3.

Boyd, Susan C. 2004. *From Witches to Crack Moms: Women, Drug Law, and Policy.* Durham, NC: Carolina Academic Press.

Cohen, Mark, Roland Rust, and Sarah Steen. 2006. "Prevention, Crime Control or Cash? Public Preferences towards Criminal Justice Spending Priorities." *Justice Quarterly* 23(3): 317–334.

Cohen, Stanley. 1972. *Folk Devils and Moral Panics: The Creation of the Mods and Rockers.* London: McGibbon and Kee.

Conrad, Peter. 1992. "Medicalization and Social Control." *Annual Review of Sociology* 18: 209–232.

Corrigan, Patrick W., Sachiko Kuwabara, and John O'Shaughnessy. 2009. "The Public Stigma of Mental Illness and Drug Addiction." *Journal of Social Work* 9(2): 139–147.

Cullen, Francis T., Bonnie S. Fisher, and Brandon K. Applegate. 2000. "Public Opinion about Punishment and Corrections." *Crime and Justice* 27: 1–79.

Fox, Renee. 1977. "The Medicalization and Demedicalization of American Society." *Daedalus* 106: 9–22.

Freed, Christopher. 2010. "Addiction Medicine and Addiction Psychiatry in America: Commonalities in the Medical Treatment of Addiction." *Contemporary Drug Problems* 37: 139–163.

Furnham, Adrian, and Louise Thomson. 1996. "Lay Theories of Heroin Addiction." *Social Science and Medicine* 43(1): 29–40.

Heyman, Gene. 2001. "Is Addiction a Chronic, Relapsing Disease?" In *Drug Addiction and Drug Policy*, edited by Philip Heymann and William Brownsberger, 81–117. Cambridge, MA: Harvard University Press.

Humphries, Drew. 1999. *Crack Mothers: Pregnancy, Drugs, and the Media.* Columbus: Ohio State University Press.

King, Ryan S. 2008. *Disparity by Geography: The War on Drugs in America's Cities.* Sentencing Project, Washington DC. Available at http://www.sentencingproject.org/Admin/Documents/publications/dp_drugarrestreport. pdf. Accessed May 29, 2008.

Lock, Eric, Jeffrey Timberlake, and Kenneth Rasinski. 2002. "Battle Fatigue: Is Public Support Waning for 'War'-Centered Drug Control Strategies?" *Crime and Delinquency* 48: 380–398.

McLellan, A. Thomas, David Lewis, Charles O'Brien, and Herbert Kleber. 2000. "Drug Dependence: A Chronic Medical Illness." *Journal of the American Medical Association* 284: 1689–1695.

Musto, David. 2002. *Drugs in America: A Documentary History.* New York: New York University Press.

National Institute on Drug Abuse (NIDA). 2006. *Principles of Drug Abuse Treatment for Criminal Justice Populations: A Research-Based Guide.* Available at https://www.drugabuse.gov/sites/default/files/podat_cj_2012. pdf. Accessed February 21, 2012.

"Ohio Survey: Many People Don't Think Addiction Is a Disease." 2010. *Alcoholism and Drug Abuse Weekly* 22(6): 7–8. Accessed via Academic Search Premier, March 2, 2011.

Peele, Stanton. 1989. *The Diseasing of America: Addiction Treatment Out of Control.* Boston: Houghton Mifflin.

Reinarman, Craig. 2005. "Addiction as Accomplishment: The Discursive Construction of Disease." *Addiction Research and Theory* 13(4): 307–320.

Reinarman, Craig, and Harry Levine, eds. 1997. *Crack in America: Demon Drugs and Social Justice.* Berkeley: University of California Press.

———. 2004. "Crack in the Rearview Mirror: Deconstructing Drug War Mythology." *Social Justice* 31(1–2): 182–199.

Rychtarik, Robert, Gerald Connors, Kurt Dermen, and Paul Stasiewicz. 2000. "Alcoholics Anonymous and the Use of Medications to Prevent Relapse: An Anonymous Survey of Member Attitudes." *Journal of Studies on Alcohol* 61: 134–138.

Smith, David, Dorothy Lee, and Leigh Dickerson Davidson. 2010. "Health Care Equality and Parity for Treatment of Addictive Disease." *Journal of Psychoactive Drugs* 42(2): 121–126.

Szasz, Thomas. 1975. *Ceremonial Chemistry.* Garden City, NY: Anchor.

Tonry, Michael. 2011. *Punishing Race: A Continuing American Dilemma.* New York: Oxford University Press.

USA Today. 2006. "*USA Today*/HBO Drug Addiction Poll." July 19. Available at http://www.usatoday.com/news/polls/2006-07-19-addiction-poll.htm. Accessed May 1, 2008.

Vastag, Brian. 2003. "Addiction Poorly Understood by Clinicians." *JAMA* 290(10): 1299–1303.

Volkow, Nora. 2007. "This Is Your Brain on Food (Interview by Kristen Leutwyler-Ozelli)." *Scientific American* 297: 84–85.

———. "Talk of the Nation." National Public Radio, February, 25, 2011.

Walley, Alexander, Julie Alperen, Debbie Cheng, Michael Botticelli, Carolyn Castro-Donlan, Jeffrey Sarnet, and Daniel Alford. 2008. "Office-Based Management of Opioid Dependence with Buprenorphine: Clinical Practices and Barriers." *Journal of General Internal Medicine* 23(9): 1393–1398.

Walters, Glenn. 1999. *The Addiction Concept: Working Hypothesis or Self-Fulfilling Prophecy?* Boston: Allyn and Bacon.

Walters, Glenn, and Alice Gilbert. 2000. "Defining Addiction: Contrasting Views of Clients and Experts." *Addiction Research* 8(3): 211–220.

DISCUSSION QUESTIONS

1. How does our understanding of "normal" affect our notions of illness?

2. What rights can a patient lose because of their condition? How does this impact the salience of medicalization?

3. If political or economic factors cause an abnormal and unhealthy situation, is it an illness?

4. What social practices could be seen as disorders or diseases?

5. How could an individual be helped if their condition was not considered a disease?

6. How could the labeling of psychosocial issues impact probability of transplants?

7. What nonmedical factors are involved in alcoholism?

8. How does the stigma of medicalization affect patients in the medical, political, and social realms?

Recommended Cases

Deafness in the Family—Parents of a deaf child are pushed to get their child a cochlear implant and are told it will limit the child's obstacles. As a low-income family, they cannot afford to maintain the implant, and the child's communicative abilities will be impacted.

Gender Dysphoria—The *DSM-5* calls gender dysphoria a "marked incongruence between one's experienced/expressed gender and their assigned gender." By maintaining the notion that transgender individuals have a mental disorder, it can increase stigma of and disenfranchisement against transgender individuals.

Nursing Homes—Rates of depression and anxiety among individuals in nursing homes are rising. Individuals who act out with increasing deviant behavior are often labeled as having mental disorders by practitioners at the nursing homes to change their status and maintain order in the facility.

CPSIA information can be obtained
at www.ICGtesting.com
Printed in the USA
LVHW101940110821
695098LV00006B/29

9 781793 522047